COMMITMENT

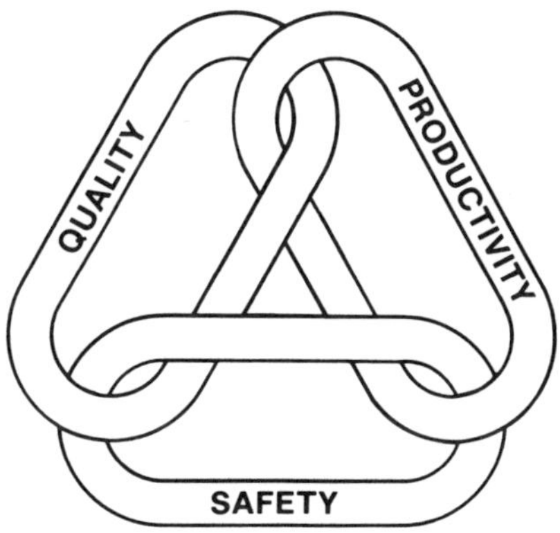

Frank E. Bird, Jr.
and
George L. Germain

Reprinted 1989
Printed in U.S.A.

Published by
Institute Publishing
(A Division of International Loss Control Institute)
Highway 78, P.O. Box 345
Loganville, Georgia 30249

Library of Congress
Catalog Card Number
(APPLIED FOR)

There is a growing awareness in business that the level of product quality required to compete (and even to survive in some industries) will not be achieved without complete employee commitment. Yet quality is often pursued in ways that neglect employee safety and health, an important aspect of quality of **people**. This leads employees to feel that management cares more about products than they do about people -- a feeling that makes it impossible to gain their full commitment to the product quality program.

This book emphasizes that the proper approach to total quality is to demonstrate a clear concern for both people and product in ways that reinforce each other — rather than having them compete for precious time and attention. The result is higher levels of safety and quality — and commitment by everyone when they see that **total quality** is actually practiced.

CONTENTS

Acknowledgement

Many quotations are used by the authors that have been taken from a variety of publications. In nearly all cases, the name of the author or source has been given with the quotation. The authors selected these with care, believing they best represent present-day opinion on the related subject. We are most certainly indebted to those named. Readers are encouraged to benefit further by obtaining one or more of the valuable related references at the close of the book.

Special acknowledgement is given to Mr. Jim Callison who provided vital professional editing of this manuscript.

Dedication

This book is dedicated to the people at First Insurance Group, Inc., San Juan, Puerto Rico, whose commitment to the prevention of loss — to people, property, process, and profit — has been a source of inspiration to the authors.

INTRODUCTION

Two vital areas of need in industry today cry out for total commitment of everyone. They can be dealt with individually, as if mutually exclusive and almost detrimental to each other, or they can be dealt with as interrelated, mutually supportive, and even dependent entities. These two specific areas of concern are quality and safety.

A widely accepted axiom used by leading experts today on quality in its broadest sense is that the foremost concern of management must be with the quality of people. Only after we, as manager, have demonstrated our concern for, and taken care of, our "humanware" can we expect their concern and loyalty for development of the "hardware" and "software" aspects of our business. This axiom provides one of the foundation stones of the Japanese approach to quality control today.

Further, in the present rush by industrialists to catch up (or even to survive) in the international market, the concentration on quality emphasis and effort may be somewhat misdirected. There is substantial evidence that worker safety, which has an enormous effect upon the way people perceive management's concern for them, may be taking a back seat to the emphasis on product quality.

In fact, many of the gains that have been made in occupational safety and health over the years could be undone if there is not a modification of the emphasis that the management of many organizations is placing exclusively on product quality.

We know that philosophically the majority of industrial leaders recognize that safety and quality are both essential. Our objective is to show the reader how commitment to both these goals can be so directed to avoid unintended misconceptions that could cause long-range damage to both concerns.

Masaaki Imai reports in his book *Kaizen* that a group of Japanese businessmen led by Toshiro Yamada, Professor Emeritus of the Faculty of Engineering at Kyoto University, recently visited industrial plants in Western nations. One of the observations made by the group concerned the apparent lack of safety precautions. A summary comment made by one member of the group was that *"There is no management if they don't care about the workers' safety."* This noteworthy conclusion by a Japanese businessman calls our attention to the assumption made by many employees of any organization when they perceive that the quality of product is getting much more attention than the safety of people.

For example, a recent newscast reported that a major automotive plant in the U.S. was cited for 82 safety violations, mostly for employee exposure to excessive levels of lead and arsenic. The total fines imposed were $1½ million. The announcer stated further that a spokesperson for OSHA said that the plant showed an almost total lack of concern for employee safety. The parent corporation of this plant has made quality the centerpiece of its advertising. The story ended with a replay of the corporate commerical about quality and worker pride; and the announcer implied there was a glaring inconsistency between the commercial and the corporate reality.

The suggestions in this book can greatly minimize the possibility of misconceptions among employees about the priorities of "hardware" versus "humanware" values in your organization. A long-range plan to implement the types of ideas expressed here will greatly increase the probability of gaining commitment of all employees to the total quality program of your organization.

Frank E. Bird, Jr.
George L. Germain

THE VENICE SUMMIT PAGE 34

INDEX FUNDS: HOW THEY'RE CHANGING THE MARKET PAGE 114

BusinessWeek

JUNE 8, 1987 A McGRAW-HILL PUBLICATION $2.00

QUALITY

To compete, U.S. companies must boost product quality. That means big changes in design, manufacturing—and management thinking. PAGE 130

SPECIAL REPORT

CHAPTER 1

Commitment — Why Is It Important?

Why is total commitment necessary for product quality and safety? Why is it so necessary to avoid misconceptions that management's concern for hardware and software is much greater than it is for "humanware"?

Explanation — Not Excuses

It isn't terribly hard to understand why some misdirection of quality emphasis could unintentionally be taken by American businessmen in particular. The vital need for product quality here was summarized by the June 8, 1987 issue of *Business Week* magazine with its cover highlighting the problem in one big word — QUALITY. Twelve pages discussing the issue can be summarized by one short sentence in an introductory article: *"Get better or get beat."* The article clearly communicated that aside from the aerospace industries and agriculture, the United States carries its own weight in very few international markets today. The once prosperous steel and textile industries are now fighting for survival. The authors of this book spent many memorable years in the steel industry and are very much aware of the pressure on this industry to deliver quality in the short term or be eliminated.

This special report in *Business Week* stated that a typical American factory invests a staggering 20% to 25% of its total operating budget on finding and fixing mistakes. In other words, as many as one-fourth of all factory personnel over a period of time are not involved with producing anything -- they are simply spending time in rework. The total cost of not doing things right the first time (hereafter referred to as "subquality") probably amounts to a punishing 30% or more of a typical company's production costs.

> *"Get better*
> *or get beat"*

The need for total commitment of all employees to quality control becomes even more evident when we consider some of the findings of a recent major survey of the American non-managerial work force by the Public Agenda Foundation:

- An amazing 50% said they put into their job **only** what is required to **keep** it.
- A majority of 75% said they **could be** significantly more effective than they are at the present time.
- Approximately 6 out of 10 admitted that they are **not** working as hard as they once did.

As a matter of fact, a more recent survey conducted by Donald L. Kanter, Chairman of Boston University's Marketing Department, and Philip H. Merves, a Professor in the school's Center for Applied Social Sciences, found that 43% of U.S. respondents were "highly cynical," held "jaundiced life views," and were "suspicious of others' motives." Of the 850 workers questioned, only 22% said they trusted management. The researchers found men were more cynical than women, and workers under 35 years of age were more cynical than older workers. The study revealed cynicism was particularly prominent in government and corporate bureaucracies. With these studies in mind, it isn't difficult to understand why we constantly hear and read about job alienation, apathy, and adversary relationships in North America; about people simply "doing their jobs" and looking to other activities for satisfaction; about lack of loyalty to employing organizations; and about a lack of pride in performance.

*" 'Subquality' amounts to
a punishing 30% or
more of a typical
company's production
costs."*

But luckily there are also some encouraging signs. For instance, a 1983 social research survey found the percentage of the American work force expressing a strong work ethic (i.e., agreeing with the statement, "I have an inner need to do the best I can regardless of pay") to be more than twice as high as in West Germany and significantly higher than in Japan. The survey also indicated the potential for better performance, revealing that 75% of American workers feel the main reason why they work less than they could is that **managers do not motivate them.**

William James of Harvard University established years ago that if motivation is low, employees' performance suffers as much (i.e., 50-70%) as if ability were low. For reasons such as this, motivating is extremely important in supervising, managing, and leading people to the point where they willingly and enthusiastically make a total commitment.

The American industrial miracle of the past hundred years was largely fueled by **capital investment** and **improved technology**. Today these are not sufficient by themselves. As stated by William J. Abernathy, Harvard professor and authority on American productivity: "I don't think industrial revitalization has much to do with investment -- it has to do with people." In other words, we need greater investment in **human resources development** and applied **behavioral science.** We need to revive and restore job pride.

Warren Bennes and Burt Nanus in their book *Leaders* point out that what really exists is a commitment gap: "Leaders have failed to instill vision, meaning, and trust in their followers."

"The attitude of foreign executives is that you can't help but succeed in the U.S. market today."

The difficulty of bridging this commitment gap as well as the importance of doing so is made clear by a comment from the issue of *Business Week* cited earlier. Mr. Armand V. Tergenbaum, President of General Systems Company and a leader in the movement toward improved quality, is quoted as warning other companies to follow the example of the favorable quality work begun in the appliance industry. He said that "offshore producers now regard virtually every U.S. market a pushover for better quality products... The attitude of foreign executives is that you can't help but succeed in the U.S. market today."

A DuPont Company executive, Ronald D. Snee, Manager of Quality Systems for Product Engineering, is quoted in the same article as saying: "The Pacific Rim countries are now doing to the Japanese what the Japanese are doing to us, and this is bound to spur the Japanese to still higher levels of quality."

貴方には出来ないでしょう

"and besides...
we know you
won't do it!"

Recently, groups of people from the United Kingdom representing management, banking, purchasing, trade union and media interests as part of a National Quality Campaign, went to Hong Kong, Japan, and the United States. Their purpose was to study quality management as related to the business success of those three economies and other competitors around the Pacific Basin.

During a rare moment of informality, one of the mission members asked a leading Japanese economist why the Japanese companies had thrown open their factories, talked so frankly to them about their techniques, and even shown so openly their secrets of success. "Because," said this Japanese leader, "it would take you ten years to get where we are now -- and by that time, we shall be even further ahead. And besides," he smiled, "we know you won't do it!"

With or without this type of challenge to a manager's ego, most American businessmen and the stock market that judges them have already accepted the fact that time is running out; and the choice is clear as reflected in an earlier comment: *"Get better or get beat."*

Many managers are off to an encouraging start and still others are gearing up through training and consulting assistance to start the accepted long march to positive change. The unfortunate reality is that when the awakening finally occurs, the problem is frequently so acute that a major consideration for many organizations is whether it's too late. In this unprecedented era of needed changes, for Americans in particular, it isn't too difficult to understand how even an item as important to the maintenance of people quality as safety could be relegated to a lower level of attention. In most cases, it probably has been given a back seat position in the hearts of those who lead. Unfortunately, safety simply has not occupied their minds enough to influence their behavior, and this frequently speaks much louder than their words. As an often-quoted Bible verse (Proverbs 23:7) points out so clearly, "as a man thinketh in his heart, so is he..." Thus, commitment to the cause(s) of his/her leaders is seriously damaged when the employee perceives that his/her personal safety and health are no longer important items on management's motivational scale.

We will not attempt to discuss here the causes or cures of the quality problem. W. Edwards Deming, Joseph M. Juran, Philip B. Crosby, and William E. Conway are among the quality experts who are well-known for their works regarding the causes of "subquality" and its prevention as a necessary business strategy. Rather, it is our purpose to stimulate some thinking about whether existing commitment in your organization is adequate and, even more important, whether your quality promotion has been directed in a manner best suited to gain the **total** commitment of everyone involved. Let's now examine more directly the need to give the subject of safety, as a major part of people quality, the proper emphasis and association with quality of product or service, in order to strengthen the total commitment of everyone toward both.

Safety Activism Factors at Work

A number of important factors remind managers today of the need for commitment to excellence in safety and health.

Power of the Consumer

Consumerism is more recent as a significant influence on safety and health management. Its goals may be clustered under a concern for "quality of life." As Alvin Toffler observed in *The Adaptive Corporation*, "Corporate retorts to the new consumerism based on traditional economic arguments cut no ice -- for the very good reason that the movement's goals are a saner, more civilized society, not necessarily a richer one."

Consumerism has emphasized growing concerns over dangers to people from manufactured products. Many products with formerly widespread use have been banned or severely restricted. Food additives, appliances, sleepwear, furniture, and even soft drink containers have all been the targets of restrictions or bans. The 1986-1987 Property Casualty Fact Book of the Insurance Information Institute reports 493 product liability verdicts in our American courts which are described as million dollar verdicts. In the case of the containers, for instance, the suspicion of birth defects arising from use of acrylonitrile in a plastic bottle resulted in the Monsanto Company's closure of four plants, the layoff of 1,000 workers, a profit reduction of fifteen million dollars, and a projected one-year sales loss of one hundred million dollars.

Unions and Collective Bargaining

In many countries, collective bargaining has played a role in workplace safety improvements for many years. Union influence has been exerted not only in direct negotiations, but also through financing or supporting of safety and health research, lobbying for safety and health legislation, and backing of liability suits filed by union members. Today's union activity in safety and health is a greater influence than ever before.

Court Awards

The editors of *Business Insurance* magazine have observed that the rise in consumerism has brought along with it an almost unrestrained tendency to sue for injury, aggravation, or even affront. In the United States, sympathetic juries have responded to these suits with astronomical awards, even though plaintiffs are sometimes motivated by greed rather than suffering. The snowballing consequences (business losses, insurance premiums, and court costs paid with tax dollars) have virtually gotten out of hand. Employees who were once constrained to be satisfied with workers' compensation settlements now sue their supervisor, plant manager, and/or safety coordinator. A guilty verdict against management personnel of a chemical company for employee injuries and death in a recent year was reported in various publications. The word "MURDER" appeared in this context for the first time in newspaper headlines. The frequency and size of awards in this relatively new area of litigation have become so enormous that liability insurance costs have risen out of sight. One article appearing recently in a national insurance journal was entitled "Sue the Bastards," and was representative of the attitude of many employees toward what they perceive to be little regard by management for their "quality of life."

Technology Increases

Technology has created a need for more extensive and dynamic safety programs. In the past 50 years, scientists and engineers have developed more new equipment and processes than were developed in the preceding thousands of years. Many tasks are becoming increasingly complex and demanding and the potential consequences of errors more costly. Also, it has been shown that virtual total safety is possible... **if** one is able and willing to pay the price in resources. The space program, for example, was highly successful, quite safe (about 99.9%), and extremely expensive. The psychological effect of these advances has been greater insistence on safety where risks had previously been assumed. People now feel that things can and should be safe. As a result, management decision-making has been broadened to encompass the potential applications of sophisticated techniques, such as a system safety analysis and a concern for the entire useful life of a product.

In brief, technological progress suggests these two vital considerations for management leaders: first, potential safety and health problems are being generated faster than ever before; and second, no problem in the workplace is too big or complex to master through professional management.

Compliance With Laws

In addition to passing the laws earlier, legislative bodies have responded to the actions of individuals, courts, and influence groups. They have started inquiries and hearings, or enacted new laws and standards where they perceived a need. For example, twenty-four hour compensation, including equal compensation for on-the-job and off-the-job accidents, is already in effect in New Zealand and is being considered in other countries. Many jurisdictions are revising workers' compensation laws to expand their application and scope.

Laws now also fix liability for disposal of materials which may at some future date affect the environment. Other laws require the study of environmental or human impact during the design and development of products. These are just a few of the legislative constraints on business which pose a real challenge to business leaders in the immediate future.

Work Force Changes

The character of the workplace has undergone great changes in the last decade. Regulations requiring employment opportunity without bias, as well as the economic need to have an added earner in the family, have brought more women into work environments which were once exclusively male precincts. Other regulations require the employment of the handicapped within their limitations. In addition, today's workers tend to be better educated and informed, and they expect more of their jobs than just a livelihood. Each of these factors introduces new demands and new challenges for management related to safety and health concerns.

Cutting Costs Without Cutting People

An additional emphasis on cost reduction stems from management's newer concern for demonstrating its concern for people in the long run as an important part of the total quality approach. Of course, cost reduction has always been one of management's key concerns. Managers are now finding that costs of premature replacement of equipment damaged or destroyed by accident are far greater than the depreciation reserves started when the old equipment was bought. Costs can be reduced if equipment can be made to last longer -- to serve at least its expected life. Consequently, accident control takes on added importance for executives.

Replacements for fire losses have even greater impact. Of businesses in North America which suffered severe fire losses in the last ten years, 43% never reopened their doors. An additional 28% failed within two years. These business failures were due to two simple factors. First, the insurance did not cover the costs of replacement: equipment, materials, and buildings had been insured at an initial (or even a depreciated) cost. Second, since insurance considerations and the safety program were separate, management had not estimated the consequences of **property damage** and **business interruption** as accidental loss.

Progressive executives and supervisors today are showing greater interest in the broad concepts of safety and health management and are working diligently to solve the problems aggravated by rampant increases in the costs of accidents of all types.

Medical Research and Toxic Substances

A relatively recent and widespread cause of concern has been the toxic substances that create health hazards in the workplace. For many years, industry has conducted research on the effects of these toxic substances and has set industrial standards, and organizations have recommended controls and exposure limits. But general lack of management response has led governmental agencies to take over the setting of standards as a logical addition to their own research and enforcement efforts. Also, medical researchers, those in governmental agencies and those working on governmental grants, are focusing their attention more and more on physical and health hazards in the workplace. Research is helping to define limits for exposures in those cases where problems can be avoided by limiting the amount or duration of exposure. More importantly, research is identifying substances and by-products which have irreversible effects and, in some cases, are carcinogenic. The growing list of diseases which may be work-related presents a greater number of liabilities possibly causing harm to workers' physical and/or mental health.

PVC is an example that most people have heard about. Until 1974, vinyl chloride gas was used in the production of over half the plastic products produced and was hardly thought of as dangerous. No one fainted or developed a rash when it was used. Then it was found to cause a rare form of cancer of the liver, which could not be diagnosed until it had reached an incurable stage. Human exposure had to be prevented. Again, management had to decide how best to control a problem discovered by medical research.

The consequences of new dangers found through medical research have sometimes been swift and drastic. Governmental control has abruptly closed down many primary product lines, dictating a total change in the objectives of the affected businesses. In many cases, research findings have also dictated expensive recalls of products when user experience has revealed hazards not acceptable to the people or the legislators who represent them. These kinds of potential consequences are among the many reasons why managers must exert the safety leadership that leads to total commitment.

An Interesting Connection

When safety is rightly regarded as a critical part of people quality and of product quality, it becomes evident that these two areas have a great deal in common. Both seek the improved performance of the items important to doing the job right the first time. Let's look at their commonalities.

Attitude:

Both areas seek to maximize the positive outlook that leads to job pride which in turn produces quality products, safely and on time.

Knowledge and Skill:

Both seek to provide adequate training and retraining as needed for critical tasks where past training has been inadequate or non-existent. The right way to do the job must include the safe way as well as the way that conforms to accepted quality requirements. It's interesting that most accidents (approximately 48%) occur during the first year of an employee's work life. Isn't it logical that this would also be the most critical period for a worker to make mistakes that result in "subquality" in the absence of adequate training? And isn't it even more interesting that research indicates that what is needed to avoid accidents is more information about how to do the job correctly -- the same need as that to achieve quality?

Fitness:

Inadequate physical or mental capability and stress have long been identified as common basic causes of accidents. The very high rate of, and management's increasing concern about, drug abuse and alcoholism on the job have been featured points of discussion in several business magazines and on the front page of the *Wall Street Journal* during the past year. Whether inadequate fitness to work results in "subquality" of work or an accident is often a fortuitous matter. The same causes are common to both, and studies show that it isn't just one main area of loss that results from inadequate fitness. It's both areas -- and the losses are big ones.

Machinery and Equipment:

Philip B. Crosby, President of the Quality Control College of Orlando, Florida, states in his book *Quality Is Free* that one of three factors causing non-conformance in an assembly area is the fact that personnel and/or tools required to perform the job are inadequate. Adequate machinery and equipment properly maintained and operated by trained personnel is unquestionably one of the biggest safeguards against occupational death and disability.

Structures:

Typically, any unsafe structure not only risks the chance of accident but work shutdown as well. Structures which make it difficult to do the job the right way can impact quality or safety, or both.

Process:

The right way to do the job (which includes the safe way) can only be consistently accomplished if the manufacturing or assembly process is also performed in accordance with standards and/or requirements. The errors that result both in "subquality" and accidents sometimes occur when performance deviations are required to compensate for things which are wrong in the process.

In summary, safety as a major part of people quality, and product quality as a part of total quality, go hand in hand. Each can reinforce the other. One without the other could eventually jeopardize the entire objective of overall quality. Care must be constantly taken to make sure our total commitment to both of these vital components of the management system is clearly visible and understood by everyone. This, in turn, makes an enormous contribution to winning the minds, hearts, and commitments of everyone. Commitment to safety **and** quality provides these benefits:

- It adds a vital dimension of purpose to the work of everyone.
- It provides self-motivation for achieving our goals.
- It creates the involvement that leads to and reinforces dedication.
- It stimulates and maintains awareness that overcomes boredom, discouragement, mental fatigue, and fear of failure.
- It is the spark for accomplishment and increased activity.
- It motivates others to want to join "the cause."

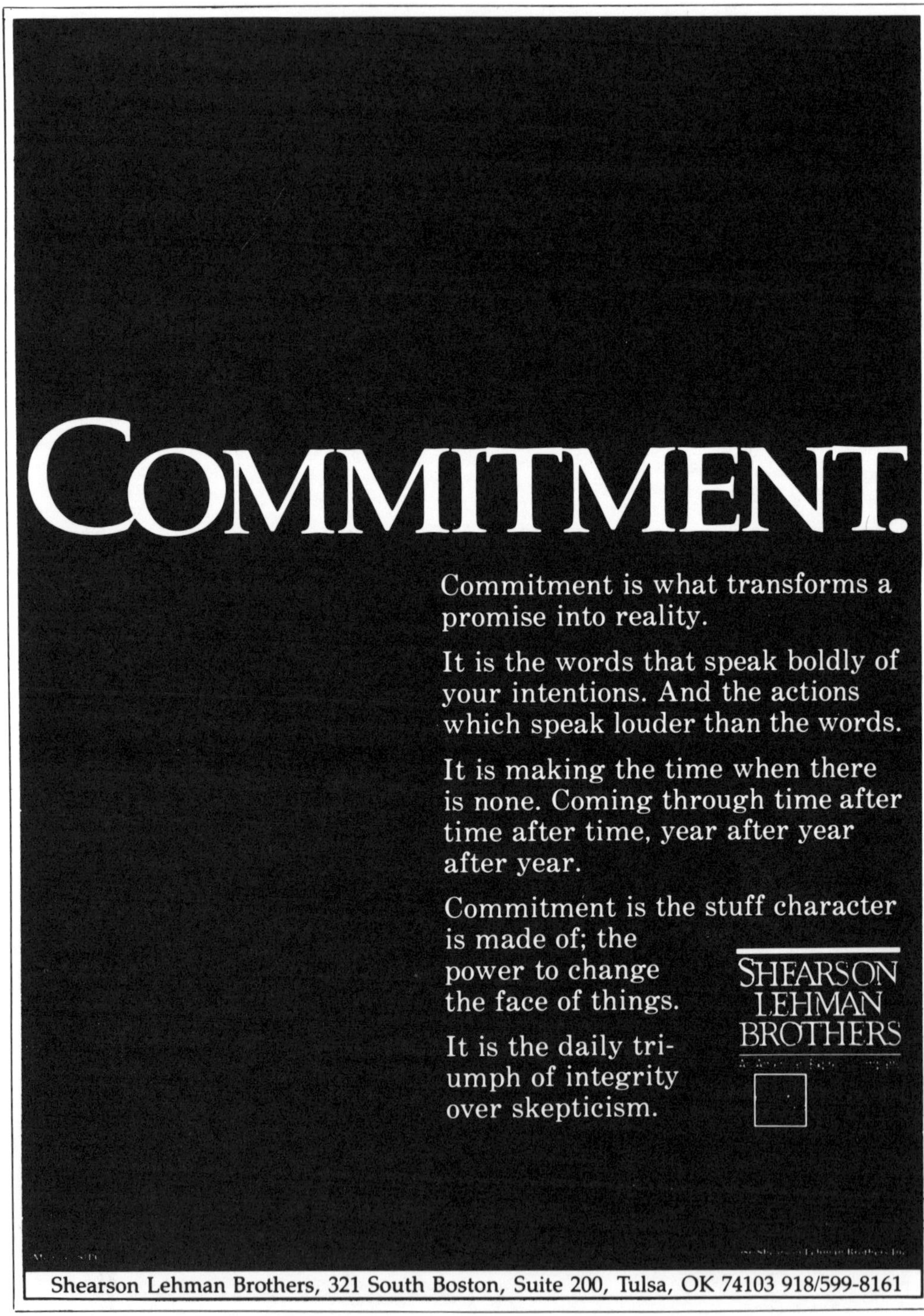
COMMITMENT.
Commitment is what transforms a promise into reality.
It is the words that speak boldly of your intentions. And the actions which speak louder than the words.
It is making the time when there is none. Coming through time after time after time, year after year after year.
Commitment is the stuff character is made of; the power to change the face of things.
It is the daily triumph of integrity over skepticism.
SHEARSON LEHMAN BROTHERS
Shearson Lehman Brothers, 321 South Boston, Suite 200, Tulsa, OK 74103 918/599-8161

CHAPTER 2

Commitment — What Is It?

As the old farmer would say, "It must be something dang important 'cause almost everyone says everybody needs it to do anything important the way they want it done."

There is hardly a management or leadership book today that does not call attention to the need for a commitment to quality. Examples include:

Michael LeBoef, in his book *The Greatest Management Principle in the World,* tells us, "Companies with a serious commitment to improving quality are gaining management support in numerous ways."

Philip B. Crosby, author of *Quality is Free,* advises that, "helping management to recognize that they must be personally committed to participating in the program raises the level of visibility for quality and ensures everyone's cooperation so long as there is some progress."

David L. Bradford and Allan R. Cohen, in their book *Managing for Excellence,* tell us that "when people feel no need to be guarded and suspicious with each other, considerable energy for task attainment is released, and a strong commitment to foster excellence is possible."

Rosabeth Moss Kanter, author of the book *The Change Masters,* points out that "when there is emotional and value commitment between person and organization, people feel that they 'belong' to a meaningful entity and can realize cherished values by their contributions."

While the most common dictionary meanings indicate commitment to be an "obligation" or "pledge," its most frequent use in safety, quality, and management literature in general usually implies an obligation or pledge of **substantial degree.** A brief review of common synonyms for the word "commitment" seems to reaffirm this belief. Words such as **given in trust, resolved, determined, converted, convinced,** and **covenanted** create an image of something that is serious and heartfelt.

The ever increasing need for commitment to safety and quality is motivating business leaders to more frequently use words like **ardor, desire, determination, devotion, faith, persistence, willpower,** and even **zeal** as they write and communicate about this important subject.

It was this type of commitment that drove people like Marie Curie, Henry Ford, Thomas Edison, Abraham Lincoln, Christopher Columbus, Albert Schweitzer, and John Kennedy to achieve extraordinary accomplishments. And yes, it is the same type of commitment that drives ordinary everyday people to overcome seemingly insurmountable barriers to success...

... in marriage and family life
... in war and acts of heroism
... in overcoming handicaps
... in athletic competition
... in devotion to their church and country.

A great illustration recently in a family publication clearly makes the point about the kind of commitment old-timers once offered when they entered an ironclad agreement by simply saying, "I give my word." We continue to quote: "Time was all that was necessary. A man's word was his bond. You could bank on it." We would do well to return to the unwavering integrity of giving one's word of commitment.

Look at this chapter in the life of a small boy who has just turned eight. At last he is able to play little league baseball. For over a year he has waited. He has a cap, a ball, a glove, and a commitment. The clock doesn't turn fast enough. The time for practice is so long in coming. He worries his mother and he calls the coach three times to be sure that practice will be held Tuesday afternoon at 5:00... Shortly after noon on Tuesday it begins to rain. By 4:30 there is a downpour. His mother says, 'Oh, son, I'm sorry it's raining so hard...' But the small boy doesn't hear. He is already out the door, cap on his head, his glove on the handlebars of his bicycle, racing toward the practice field on the school ground. The coach, who lives across the street, looks through the window; it's raining too hard to practice. But to his surprise he sees in the downpour a small boy standing where home plate once was, up to his knees in water. He grabs his raincoat and rushes out to rescue the boy. 'Can't you see it's too wet to practice?' The little fellow answers, 'But, Coach, I told you I'd be here.' In time, by the example of his elders, the boy may learn 'the way things are done.' But right now all he has is his integrity — he just gives his word, the word of commitment — and sticks to it. With him, it all begins and ends with commitment."

One characteristic of a professional leader is that he/she uses a functional vocabulary common to that used by colleagues with meanings that may differ somewhat from those found in dictionaries. The word "commitment" has a functional meaning that would imply an obligation or pledge of substantial degree, similar to the kind exhibited by the lad in the earlier illustration. To prove this, we pulled a surprise on colleagues who were recently with us in Atlanta. Without giving them warning or opportunity to prepare, we requested that they promptly write a definition of commitment as related to what leaders seek from employees in safety and product quality.

Hugh Douglas, a respected Canadian professional and author of several fine books on safety management, wrote:

> *"Total willingness to unwaveringly apply all necessary resources to accomplish the objective or goal."*

Dr. M. Douglas Clark, a great contributor to safety literature and a professional leader on our Institute staff, contributed:

> *"The willingness to pay the necessary price to achieve a desired objective."*

Others also jotted their immediate thoughts on paper:

> *"The highest level of attitude development toward a subject where one's consistent actions and words communicate a deep belief and devotion to its cause."*

> *"Total willingness to persistently devote all required resources to accomplishment of a mission."*

The type of commitment described by any of these definitions was exhibited recently by an employee of an industrial plant who was attending a fund-raising campaign meeting for a youth center in a neighboring community. He walked into the meeting room, sat down, and immediately introduced himself. Reaching out his hand with a broad smile he said, "How do you do? I am Charles Downing, a pipe fitter at that large plant you probably saw if you came in from the west end of town. You know," he went on to say as if there were a built-in automatic recording, "our department has held the President's Honor Efficiency Award which includes safety, quality and overall work performance in our company, for over two years. And you know," he continued, "we plan to keep it for many more. What's your name, sir, and who do you represent?" This unsolicited comment at a public meeting and the attitude it represented gave enormous evidence that this individual had the spirit of commitment that many managers hope they can instill in their people.

The following pages will show how hundreds of progressive managers are successfully guiding their people through the stages of attitude development leading to the commitment required by today's vital safety and quality needs.

The Route to Total Commitment and the Reinforcement Cycles That Maintain It

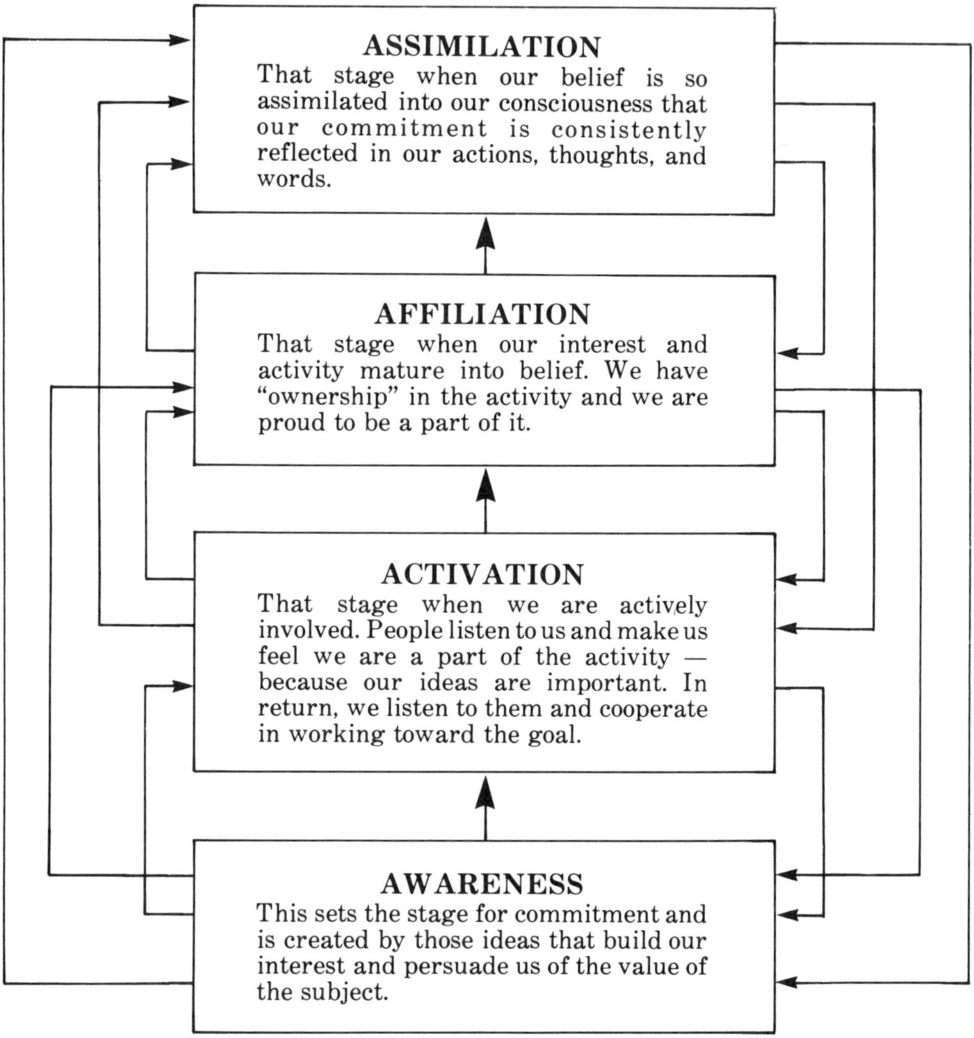

Constant reinforcement is essential through involvement at all stages to maintain one's existing level of growth and development.

CHAPTER 3

Commitment — How Do You Develop It In Others?

Michael LeBoeuf, author of *The Greatest Management Principle in the World*, gives enormous insight into motivating loyalty and commitment from people. He instructs his readers to "find people who have been with a firm at least ten years and are still very excited and dedicated to the company. Then," he continues, "ask them why they take so much pride and feel so much loyalty to the company." He points out that "you will most likely get a number of answers, but they will boil down to this: they belong to an organization that cares about them, challenges them, believes in them, and wants the best for them, not just as employees but as total human beings."

Howard S. Gitlow and Shelly J. Gitlow confirm this in *The Deming Guide to Quality and Competitive Position* with this observation: "Employees must be viewed as the most valuable resource of the company and their pride as essential to the company's existence in the long run. Ignoring the needs of the work force is extremely shortsighted because it ignores the benefits of having healthy, motivated workers."

"You can buy a man's time; you can buy his physical presence at a given place; you can even buy a measured number of his skilled muscular motions per hour. But you cannot buy enthusiasm... You cannot buy loyalty... You cannot buy the devotion of hearts, minds, or souls. You must earn these."

—Clarence Francis

The suggestions that follow will indicate the authors' strong agreement with these excellent comments about commitment. Because of our conviction, we extended an earlier caveat that in the present rush to catch up and even to survive in the international market by many industries, the concentration of quality emphasis and effort may be somewhat misdirected. We suggested that the evidence indicates that worker safety, which has an enormous effect upon the way people perceive management's concern for them, may be taking a back seat to the emphasis on product or service quality. At the same time, by occasionally giving safety efforts an exclusive emphasis and by not demonstrating through behavior that safety is an important part of doing things the right way, management frequently does a great injustice to the significant values which could be contributed by the program.

One of the major reasons for writing this book was to demonstrate as clearly as possible how a properly directed safety and quality emphasis in an organization could increase the probability of total commitment by everyone to both of these vital areas.

"The most valuable fringe benefit of considering safety an operational strategy has been the general improvement in line management's ability to manage all aspects of their total jobs."

—Lester A. Hudson
President and
Chief Operating Officer
Dan River, Inc.

An article entitled "Delegation for Employee Development," by Donna Venton, published in the January, 1987 issue of the *Training and Development Journal*, included a summary of one of the few studies ever conducted to identify the causes of commitment. That summary is quoted below:

Developing Organization Commitment

How commitment to the organization is developed and maintained is of central importance to organizations and has no easy solutions nor hard and fast answers. But in a 1975 study of 279 managers from eight large U.S. companies -- three of which were Fortune 500 manufacturing concerns and five of which were federal government agencies -- Bruce Buchanan attempted to identify the causes of commitment. The findings identified experiences that fostered commitment among all 279 managers; then multiple regression analysis was used to identify those experiences having the strongest association with the development of commitment.

(continued)

Developing Organization Commitment (continued)

The five classes of experience having a significant impact on commitment, in order of strength, were:

- a feeling of personal importance from being considered productive and valuable to the organization
- experience in a cohesive group with positive feelings toward the organization
- realization of expectations (the question asked was, "Has my organization fulfilled its promises to me and otherwise met my expectations in areas I care about?")
- organizational commitment norms (if employees who thought that their organizations expected them to be committed were more committed than those who didn't sense those expectations)
- first-year job challenge.

Buchanan notes that the impact of the last item was both predictable and surprising: predictable in that "a challenging, interesting, and self-confirming work assignment would strengthen commitment," and surprising in that "the job experience of the first year continued to shape the attitudes of managers."

While the study highlighted experiences that fostered commitment among 279 management people, the findings could apply equally well to non-managerial personnel.

Since little specific data of a similar nature has been published, this summary, along with the two quotations at the beginning of this section, might verify that the suggestions which follow will logically help to achieve the desired commitment.

Listed on the following pages are certain activities important to success in occupational safety and health. While usually coordinated by safety and health professionals, they can also have an enormous direct and/or synergistic value for quality if properly directed. While they are important, they are purely representative of the total activities that would constitute the entire organization's effort. Each activity will be followed by the suggested **potential problem** and **potential solution.** A **potential problem** should be interpreted in this case as a circumstance or situation that could lead to misunderstanding and ultimately cynicism by employees about management's true concern for them -- in effect, roadblocks to total commitment. Please recognize that the ideas are suggestive, designed to prime your mental pump to use innovative and creative thinking needed to accomplish your total commitment goal.

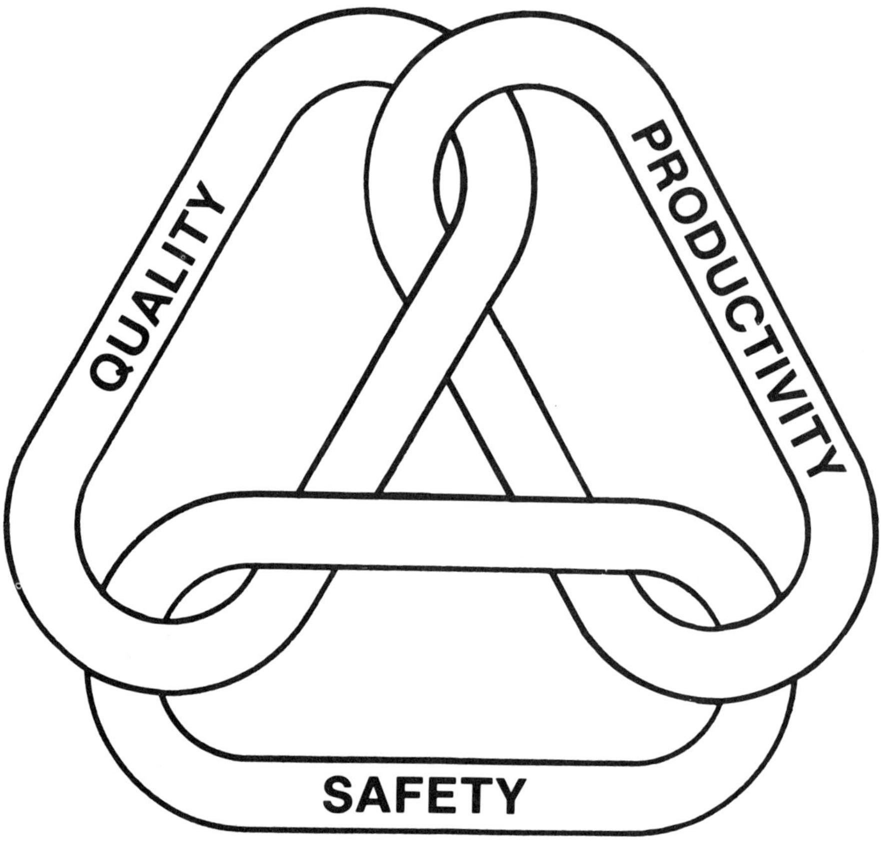
QUALITY
PRODUCTIVITY
SAFETY

Activity 1 — Mission Statement

Potential Problem

The mission statement is usually a clear statement of the organization's mission incorporating the major values that undergird the mission's accomplishment. Invariably, progressive companies today express the value of people in terms of a strong concern for safety, health, and environmental matters as a primary need with the value of customer focus and product quality usually following in its implied importance to the organization. Most mission statements clearly establish the principal role of the organization, and without question, a primary concern and recognition of the value of its people is emphasized above other key values. **The potential problem here is not a well-written statement that people want to hear. It's the hypocrisy on management's part that is perceived when its behavior seems to contradict its stated mission.**

Potential Solution

On many occasions when key management personnel engage in behavior(s) related to quality, they could easily indicate with sincerity that the item of emphasis is just as beneficial to safety. Forming the new habit of highlighting SQP (Safety/Quality/Productivity), managers should emphasize "the right way" instead of product quality as a **mutually exclusive** priority. Likewise, the reverse is also true: safety is too often presented as a separate entity and not something that should be integrated into, rather than separated from, the total actions of everyone. Thus, an opportunity to maximize the potential benefits to quality and productivity, as well as safety, has been missed.

Activity 2 — Policy Statement

Potential Problem

The safety and health policy is frequently typewritten and posted with other personnel notices on the regular company bulletin boards. The quality policy, on the other hand, is usually professionally typeset in multicolor printed form, widely posted, and mounted in attractive frames. The message as to what is really important begins to take shape — and the message is that it isn't the safety of people.

Potential Solution

Loss is avoidable waste of any resource. Instead of writing several different policies on such subjects as "safety and health," "environmental control" and "quality," why not write **one** on "Resource Protection"? By doing so, we will minimize the possibility of misinterpretation of our mission, with the resulting cynicism as to what management really wants. It suggests that **all** programs would benefit from less cost and confusion. Avoid behavior that emphasizes one policy and tends to communicate that it deserves the most attention. Write and print policy statements (for safety and quality) together in similar form. Comment freely that SQP (Safety, Quality, and Productivity) go hand in hand.

Activity 3 — Upper Management Visibility

Potential Problem

Actions during field tours, in written communications, and in meetings sometimes are perceived as giving emphasis to one item over another. Particularly if that item is quality, productivity or costs with the exclusion of safety, (even when in actuality that was not what was intended), that kind of behavior can cause morale problems among employees. "I hear what you say, but I see what you mean" is the obvious conclusion of most people.

Potential Solution

Think SQP as we communicate and "manage by walking around." This doesn't mean that we shouldn't address individual problems in any singular activity of management. It does mean that if the particular point of our communication would be beneficial to other SQP areas, we ought to make it an automatic reflex to say so. Get in the habit of using all senses to recognize that most of the time the importance of "the right way" to do things is the message we want to convey.

Activity 4 — Identification of Program Activities and Performance Standards

Potential Problem

Identifying program/process activities as mutually exclusive to safety or quality (i.e., Job Safety Analysis as the safe way to do the job, Job Quality Analysis as the way to do it for quality, resulting in a Job Safety or a Job Quality Procedure, Planned Job/Task Safety Observation, etc.) is another example of misplaced emphasis. Not only does this exclusive approach create additional paperwork and cost by the repetitious training and involvement necessary, but it can actually be counterproductive to its intended objective of of giving people adequate information to do the job right. Common sense as well as research tell us that accidents and "subquality" occur because people do not have enough information about doing their jobs properly. The safe way or the quality way in exclusivity to the other can defeat its very purpose. The particular safe way selected may not be suitable to quality or productivity; and likewise, safety may be compromised when the quality procedure is exclusive in its direction. We are fully aware that a number of major corporations are already generating a voluminous number of Job Quality Analyses and Procedures. This is particularly noted in those companies where safety has been dealt with as mutually exclusive. It isn't difficult to imagine the frustration that is bound to come, especially from the front-line supervisors and non-managerial employees, as all other quality program activities take the same route.

Potential Solution

Job/task safety and quality analyses should be an important part of a complete job/task analysis and should result in what the authors prefer to call the standard job/task procedure. It's important to emphasize the need for safety and quality analyses. Our position is that the resulting procedure, whatever we choose to call it, should incorporate the causes of loss to the system: people/equipment/material/product. In effect, the end product is really "the right way to do the job."

This basic philosophy would relate to many other educational and motivational program/process activities (i.e., planned job/task observation should not have exclusive direction toward safety or quality but should be directed at the total job/task and **the right way to do the job**). It is surprising how many barriers to interdepartmental cooperation have been created by the isolation in which many organizational groups have worked in recent years. One other example will illustrate the potential mutual values: incident investigation forms can be used for "subquality" incidents **as well as accidents for near-miss events in either important area.**

Effective Communication and Motivation Tools to Encourage and Commit People to Do Their Work the Right Way:

This	Not This
Proper Job/Task Instruction	Job/Task Safety Instruction Job/Task Quality Instruction
Standard Job/Task Procedures	Job/Task Safety Procedures Job/Task Quality Procedures
Planned Job/Task Observations	Job/Task Safety Observations Job/Task Quality Observations
Positive Behavior Reinforcement for the Right Way to Do the Job	Safe Behavior Reinforcement Quality Behavior Reinforcement

EXHIBIT 1

Safety and quality professionals must draw themselves closer together during this period requiring the best "change management" skill of everyone. Not long ago, a safety manager related that he has tried to stimulate a cooperative relationship with quality control personnel in one of the world's largest companies; in an industry acknowledged to be totally outclassed by the Japanese quality effort. "It's virtually impossible for such relationships to develop," he stated. "They are running so scared in protecting their own turf and jobs that cooperation with us is out of the question." The Japanese have been taught well by both W. E. Deming and J. M. Juran. They know that commitment to quality will only come by "driving out fear" and "breaking down barriers between staff areas." That is why a leading Japanese economist was quoted earlier as saying to English-speaking visitors, "and besides, you won't do it."

The authors obviously feel otherwise, and that is why we were motivated to develop these thoughts. We know that fear and barriers to cooperation and commitment **can** be destroyed. This is confirmed by an ever increasing number of progressive organizations which are growing in their commitment day by day.

Activity 5 — Measurement and Evaluation System to Determine Compliance With Performance Standards at All Levels of Organization

Potential Problem

Lack of interdepartmental cooperation in the development of measurement techniques that are not mutually exclusive to safety or quality interests makes total SQP performance difficult to evaluate. Certain specific statistical controls and other measures of performance will need to be exclusive to either important area of management's interest. However, there are many more that are common in their contribution to "the right way to do the job" that are often developed independently and exclusively in the interest of safety or quality. Supervisory and non-managerial people alike get frustrated by what is accurately perceived to be the overlapping and duplication in similar programs requiring extra time, effort, and a never-ending volume of paperwork.

Potential Solution

Closer cooperation between safety and quality personnel is needed for them to know and understand the measurement and evaluation systems being utilized in both programs, and whenever practical, to use each other's systems to mutual advantage. One of the most important program evaluation systems deserving priority attention is the periodic, comprehensive audit. Safety personnel have long utilized this valuable tool that only in recent times has been gaining wide application in quality programs. These audits can involve a significant number of days to complete and may require the time of many key people in an organization. Many modern safety audit systems measure and evaluate a significant number of program activities that are vital to the control of quality in the organization. This could include numerous communication and motivation programs, such as job orientation, personal contacts, standard job/task procedures, planned job/task observation, and supervisory and employee training. Improved cooperation would not only minimize duplication and wasted time and effort by many people, but could, through sharing, assist in developing stronger programs for both.

Activity 6 — Adequate Training at All Levels

Potential Problem

Managerial and non-managerial training needs are not properly evaluated. Art Resnick, girls' high school soccer coach, was selected as Coach of the Year by the National Soccer Coach Association in 1983. A simple comment made by Coach Resnick was widely utilized by business editors in such publications as *The Boardroom Reports, The Wall Street Journal,* and the business section of *The New York Times.* This highly motivational comment was simple: "Practice doesn't make perfect -- it makes permanent." Coach Resnick was simply stating the power of habit formation. If we don't start workers off with the proper orientation or give them the right instruction whenever it is necessary, their practice of that improper orientation, that incorrect instruction, will not make their performance perfect no matter how many times they do that job; but it will make it permanent.

Since training needs have not been properly evaluated in a high percentage of organizations, inadequately trained employees frequently learn the wrong way from the start or at critical stages of their development. Not only is the training a costly and sometimes unnecessarily hazardous one, but it is also a stressful one for the worker as well as those who interact with him/her. Much of the time the need for adequate training to avoid accidents, poor quality, and general waste is obvious to everyone except those who make the decision to have it done. It's not hard to imagine the difficulty in gaining any type of commitment in an atmosphere where people have inadequate knowledge and skill to do their jobs safely or well.

Several years ago, Stephen A. Sawzin compared the cost of structured versus non-structured training for semi-skilled production workers. He measured the relative economic return in terms of competence of the trained workers using performance data, product quality, raw material, and worker attitudes. The structured training was a well-organized, systems approach, learner-centered program with specific job competencies as training objectives; while the unstructured training was the typical laissez-faire situation where the trainees learned on the job through their own initiatives and unorganized coaching by fellow workers.

Mr. Sawzin found that unstructured training required 250% more time to achieve job competency, and had a 230% loss in production rate and a 316% increase in raw material waste.

One of the most important training needs for many organizations is adequate orientation of new workers. This need is verified by a recent U.S. Department of Labor statistical analysis showing that 48% of workers sustaining injuries (the kind required to be reported to the federal government) were injured during their first year of employment. Whenever we think of new employees, we must always keep in mind the transferred worker. There is enormous evidence to support the fact that the basic causes of loss, whether accident or poor quality, are the same (see Exhibit 2). It follows that if inadequate training is given to new employees, workers will be as prone to make the mistakes that cause "subquality" as those that result in accidents.

Potential Solution

Do an effective training needs evaluation for any job/task where there is a history of, or potential for, major safety, quality, or productivity problems. Performance problem analysis techniques have long been used by profesional trainers to systematically identify training needs. Related techniques that help pinpoint training needs could include work analysis, job observation, and tests and surveys of different types. These methods have been utilized by many progressive safety leaders in identifying training needs for safety. By breaking down the barriers to interdepartmental cooperation, personnel in safety and quality can more effectively work together in their programs to satisfy training needs frequently common to both.

Loss Causation Principles

1. The basic causes of accidents and "subquality" are the same.
2. Chance determines whether these common causes result in loss to people, property, or product.
3. The vast majority (80-85%) of the accidents or "subquality" causes result from inadequacies in the management system.
4. Losses to people, property, and products don't "just happen"; they are caused.
5. The basic causes of loss can be effectively controlled by an adequate program, program standards, and compliance with those standards.

EXHIBIT 2

As stated earlier, quality control personnel are introducing job/task quality analyses and procedures, thus doubling management efforts and paperwork already required by job/task safety analyses. There is also a doubling of training efforts in a significant number of other program activities by safety and quality personnel. This problem will continue to grow unless a spirit of cooperation, not presently perceived in most organizations, evolves in the near future. Deming, Juran, and Crosby all emphasize training as a vital need of an effective quality program. Training was also one of the five most needed activities revealed in safety program evaluations by auditors from over 200 companies in 1987. Imagine the efficiency and cost reduction potential of an all-out cooperative training effort by safety and quality personnel.

Activity 7 — Organized System of Participation at All Levels

Potential Problem

Attempting to implement a safety or quality control activity without maximizing the involvement of people at all levels of the organization leads to lack of support and commitment. The problem is further amplified when, as indicated earlier, the activity concerned is being utilized as exclusive to safety or quality when it could benefit both. When the effort is also narrow in its inclusion of the "right people" at the planning and implementation stages, it is seen as their idea or program rather than "ours." One of the important basic truths in **change management** is that "people support what they help create."

Potential Solution

Manage the direct involvement of all members of management, and direct or indirect involvement of as many non-managerial employees as possible, in implementation activities.

An increasing number of organizations are utilizing a team system that gets all members of management directly involved in implementation of an important segment of a program at some point during a 2-3 year period. To assist directly in implementing a facet of the program, it is necessary to know how other aspects interrelate. Thus, all participants must know and understand the entire program. Implementation team activity is planned by a subgroup of middle managers chaired by someone from upper management. The team, made up of personnel down to and including front-line supervisors, functions for a relatively short period of time to implement or reinforce a facet of the program. If there is close cooperation between safety and quality personnel, many facets of either the safety or quality emphasis can receive mutual benefits. One of the key responsibilities of a team chairperson is to evaluate the attitudes of all members of the implementation team and to do everything possible to build or reinforce commitment during the time that particular team is active.

One method is to convert attitudes to behavior, so the level of commitment by team members can be evaluated by the team leader. One way of doing this is to give a test in order to find out how they feel and think about safety and quality, by having them write down their ideas. Another way would be to have members verbally respond to questions presented orally. The disadvantage of this is that people tend to be influenced by what others say or the body language of the leader. An interesting phenomenon is that the person who is attempting to increase the commitment of others is reinforcing his/her own in the process.

There is a useful quotation attributed to a committee appointed by Dean Armond Drake at a major psychological clinic at Columbia University several years ago. It summarizes the mental process involved with attitude development that leads to action:

> "What the mind attends to it considers;
> What the mind does not attend to it dismisses;
> What the mind attends to regularly it believes;
> And what the mind believes it eventually does."

There seems to be no doubt in the minds of psychological experts that involvement leads to belief; and belief leads to action which, when properly reinforced, inevitably leads to commitment.

To gain as much involvement as possible at all levels, a wide variety of methods must be employed. These can include, but are not limited to, safety or quality project team assignments, critical job/task inventory team assignments, group meetings, personal contacts, suggestion systems, condition report systems, and SQP (Safety, Quality, Productivity) circles. Closer cooperation between safety and quality personnel could result in activities by both that would incorporate features of mutual value. These types of program activities with values common to safety and quality represent tremendous potential for cooperative effort to not only minimize wasted time and effort but to also increase program acceptance and participation by everyone.

Even more important, closer association between safety and quality activities of an organization gives clearer indication to everyone that people are management's most important resource. Generating this feeling in American employees in particular would appear to be a critical need today, as indicated by the studies cited earlier in this book. American managers sometimes forget that the Japanese cultural environment into which Deming and Juran's efforts gained such enormous acceptance were, by their nature, fertile ground for quality emphasis. The Japanese have been masters at gaining commitment and dedication for centuries. One has only to remember the kamikaze pilots of World War II who volunteered by the thousands to give their lives in suicidal missions. Historians point out that during the final days of the war only a dramatic appeal and proclamation by the Emperor of Japan prevented millions of people from commiting suicide. They would have taken their own lives willingly, not from fear but from belief and commitment to an act they believed essential to their religious and historical pride.

In modern Japan that commitment, influenced to a great extent by the effects of ancient social and religious influences, has faded. Within industry, however, a new type of commitment flourished in the fertile soil of this interesting culture. Many employers have committed themselves to a "cradle to grave" responsibility for their people. This, combined with historical and cultural influences, creates an unbelievable willingness in the minds and hearts of Japanese workers to give commitment in return, so long as effective leadership on the job supports these strong cultural influences.

A number of respected consultants and writers on leadership skills and techniques have said that gaining commitment can be summarized by these few simple words: *"you get commitment by giving commitment."*

Delta Airlines, known as one of the most successful U.S. companies today, has found the elusive program ingredients that generate employee loyalty and commitment. Its efforts in maintaining a well-run, profitable organization with a loyal work force have been cited in a number of prominent business publications.

A popular TV advertisement shows a Delta employee carefully placing baggage on a conveyor as he comments: "I'm proud of my job and I'm proud of Delta. When you feel that way, you just naturally take care." The authors, who fly regularly, have sampled the attitudes of Delta employees by asking several of them, "What makes Delta a great company?" While the TV commercials have probably been seen by most Delta employees and they may very well receive regular coaching in public relations, these employees' attitudes seem genuine nevertheless. Their responses are natural, sincere and diversely stated, yet all convey the same message. Said the driver of an electric cart recently, summarizing many other comments given, "We work for a company that cares about us, and we know we'll have our jobs in the future."

Developing the type of attitude that reflects a significant depth of commitment by everyone in an organization doesn't come easy -- and it doesn't come overnight. Philip B. Crosby, in his book *Quality Is Free,* states that "it takes four or five years to get people to understand the need for, and to learn to have confidence in, such an improvement program."

Peters and Waterman suggested in 1982 that it takes five to ten years to get an entire large organization to accept a new superordinate goal. Many years of experience working with several hundred companies in safety and health program implementation has shown that it takes five to ten years for an average company to achieve the level of performance of its leaders.

The achievement of total commitment takes an enormous amount of time and persistence to ultimately reach that level in employee attitude where it can be said, "We have it." Motivational leaders who have written about success have repeated this thought hundreds of times:

Ben Sweetland states, *"Success is a journey, not a destination."*

Dr. Robert Schuller, the well-known television preacher and author of several popular books on success, points out that *"inch by inch, success is a cinch."*

Anthony Greenwood, Ohio State University psychologist, suggests that *"all you need to succeed is to keep on trying."*

The great coach Vince Lombardi said something quite similar with a little modification: *"The difference between a successful person and others is not a lack of strength, not a lack of knowledge, but rather a lack of will."* He constantly guided his players with the statement, *"There are only three things that are important in your life: (1) your God, (2) your family, and (3) the Green Bay Packers."*

As presented in detail elsewhere (Bird and Germain, "Job Pride Development" chapter of *Practical Loss Control Leadership*), many of these motivational dynamics can be summarized in terms of these four powerful success factors:

1. GOAL POWER: Motivation to accomplish results tends to increase when people have meaningful goals toward which to work.
2. PARTICIPATION: Motivation to accomplish results tends to increase as people participate in matters affecting those results.
3. FEEDBACK: Motivation to accomplish results tends to increase as people are informed about matters affecting those results.
4. RECOGNITION: Motivation to accomplish results tends to increase as people receive recognition for their contribution to those results.

The thoughts in this book can assist managers in their efforts to achieve success by gaining the commitment of their people. These suggestions are related to safety and quality program implementation methodology because of the strong impact we believe this area has on employee attitude toward total commitment. This emphasis was influenced by our perception that it is not given the attention it deserves in the literature devoted to development of job pride within the worker. The authors certainly agree there are multiple causes involved with attitude and commitment development that deserve our continuous study. We hope nevertheless that some of the thoughts expressed here will assist the reader in his/her efforts to gain what has been an elusive goal. We have everything to gain by doing those things that can win people over to our commitment cause. The only things we stand to lose are accidents and "subquality" work.

Accident and "Subquality" Loss Causation Model

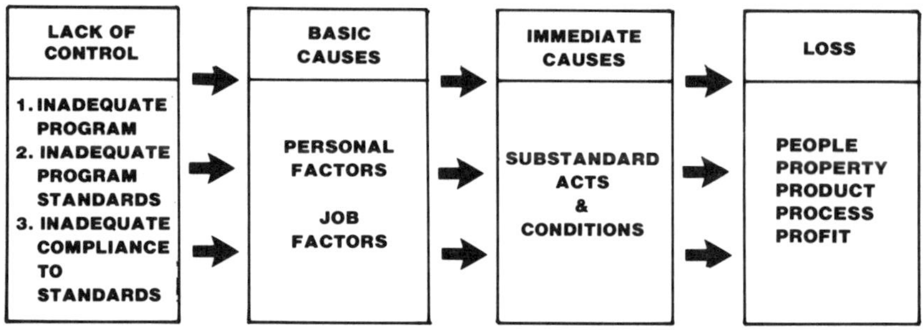

Loss is any avoidable waste of a resource. Loss Control Management is the application of professional management skills and techniques to the control of that avoidable waste. By controlling the adequacy of the safety or quality program, its standards, and the compliance to those standards, within levels of acceptable limits, the management team can prevent or control unacceptable deviations that result in loss. **Accidents and "subquality" can be prevented or controlled by maintaining adequate performance standards and measuring compliance with those standards.**

The management system must be ever alert to the need for change as it works to maintain and improve this dynamic system by planning, organizing, leading, and controlling with the entire employee team.

The basic causes of loss, which are common to accidents and "subquality," will appear in the system if control is not maintained effectively. These basic causes will lead to the substandard practices and conditions that come together under certain circumstances to produce losses to people, property, product, process, and/or profit.

EXHIBIT 3

Basic Causes of Accidents and "Subquality"

PERSONAL FACTORS

- **Inadequate Physical/Physiological Capability**
  - inappropriate height, weight, size, strength, reach, etc.
  - restricted range of body movement
  - limited ability to sustain body positions
  - substance sensitivities or allergies
  - sensitivities to sensory extremes (temperature, sound, etc.)
  - vision deficiency
  - hearing deficiency
  - other sensory deficiency (touch, taste, smell, balance)
  - respiratory incapacity
  - other permanent physical disabilities
  - temporary disabilities

- **Inadequate Mental/Psychological Capability**
  - tears and phobias
  - emotional disturbance
  - mental illness
  - intelligence level
  - inability to comprehend
  - poor judgment
  - poor coordination
  - slow reaction time
  - low mechanical aptitude
  - low learning aptitude
  - memory failure

- **Physical or Physiological Stress**
  - injury or illness
  - fatigue due to task load or duration
  - fatigue due to lack of rest
  - fatigue due to sensory overload
  - exposure to health hazards
  - exposure to temperature extremes
  - oxygen deficiency
  - atmospheric pressure variation
  - constrained movement
  - blood sugar insufficiency
  - drugs

- **Mental or Psychological Stress**
  - emotional overload
  - fatigue due to mental task load or speed
  - extreme judgment/decision demands
  - routine, monotony, demand for uneventful vigilance
  - extreme concentration/perception demands
  - "meaningless" or "degrading" activities
  - confusing directions
  - conflicting demands
  - preoccupation with problems
  - frustration
  - mental illness

- **Lack of Knowledge**
  - lack of experience
  - inadequate orientation
  - inadequate initial training
  - inadequate update training
  - misunderstood directions

- **Lack of Skill**
  - inadequate initial instruction
  - inadequate practice
  - infrequent performance
  - lack of coaching

- **Improper Motivation**
  - improper performance is rewarding
  - proper performance is punishing
  - lack of incentives
  - excessive frustration
  - inappropriate aggression
  - improper attempt to save time or effort
  - improper attempt to avoid discomfort
  - improper attempt to gain attention
  - inappropriate peer pressure
  - improper supervisory example
  - inadequate performance feedback
  - inadequate reinforcement of proper behavior
  - improper production incentives

EXHIBIT 4

Basic Causes of Accidents and "Subquality"

JOB FACTORS

• Inadequate Leadership and Supervision
- unclear or conflicting reporting relationships
- unclear or conflicting assignment of responsibility
- improper or insufficient delegation
- giving inadequate policy, procedure, practices, or guidelines
- giving objectives, goals, or standards that conflict
- inadequate work planning or programming
- inadequate instructions, orientation, and/or training
- providing inadequate reference documents, directives, and guidance publications
- inadequate identification and evaluation of loss exposures
- lack of supervisory/management job knowledge
- inadequate matching of individual qualifications and job/task requirements
- inadequate performance measurement and evaluation
- inadequate or incorrect performance feedback

• Inadequate Engineering
- inadequate assessment of loss exposures
- inadequate consideration of human factors/ ergonomics
- inadequate standards, specifications, and/or design criteria
- inadequate monitoring of construction
- inadequate assessment of operational readiness
- inadequate monitoring of initial operation
- inadequate evaluation of changes

• Inadequate Purchasing
- inadequate specifications on requisitions
- inadequate research on materials/equipment
- inadequate specifications to vendors
- inadequate mode or route of shipment
- inadequate receiving inspection and acceptance
- inadequate communication of safety and health data
- improper handling of materials
- improper storage of materials
- improper transporting of materials
- inadequate identification of hazardous items
- improper salvage and/or waste disposal

• Inadequate Maintenance
- inadequate preventive
  ...assessment of needs
  ...lubrication and servicing
  ...adjustment/assembly
  ...cleaning or resurfacing
- inadequate reparative
  ...communication of needs
  ...scheduling of work
  ...examination of units
  ...part substitution

• Inadequate Tools and Equipment
- inadequate assessment of needs and risks
- inadequate human factors/ergonomics consideration
- inadequate standards or specifications
- inadequate availability
- inadequate adjustment/repair/maintenance
- inadequate salvage and reclamation
- inadequate removal and replacement of unsuitable items

• Inadequate Work Standards
- inadequate development of standards
- inadequate communication of standards
- inadequate maintenance of standards

• Wear and Tear
- inadequate planning of use
- improper extension of service life
- inadequate inspection and/or monitoring
- improper loading or rate of use
- inadequate maintenance
- use by unqualified or untrained people
- use for wrong purpose

• Abuse or Misuse
- condoned by supervision
  ...intentional
  ...unintentional
- not condoned by supervision
  ...intentional
  ...unintentional

EXHIBIT 4 (Continued)

CHAPTER 4

Commitment — How Do You Develop It In Yourself?

The authors of this book have worked together for well over twenty years and have jointly collaborated on a number of books and other projects. We continue to refer to SQP with the idea that Safety, Quality, and Productivity go hand in hand. One of them is not "the right way" to do the job without the others.

> "Victories do not always go
> To the stronger or faster man.
> Sooner or later the one who wins
> Is the person who **believes** he can!"

The idea of "the magic of believing" and the power that comes to the individual who uses it is frequently referred to as the power of "mind over matter" ... the influence of our attitudes on our actions... the effects of our beliefs on our behavior. This idea's importance has been reflected in what we have discussed in the earlier chapters of this book.

When Dr. Roger Bannister first broke the 4-minute mile barrier, he said it was more the result of **mental** conditioning than **physical** conditioning. He had to make up his mind, to **believe** that he could break the record, before his mind would let his body do it. Believing and achieving go hand in hand; therefore...

BELIEVING IN SQP EXCELLENCE
AND ACHIEVING SQP EXCELLENCE
GO HAND IN HAND.

Examples of this power of believing come from all fields, such as sports, religion, medicine, science, government, and business. Examples include:

- Many people who truly BELIEVE in the power of prayer are relieved of afflictions as if by magic.
- Doctors attest to the healing power of attitudes/willpower/BELIEF. They know this often makes the difference between life and death or health and sickness.
- Franklin D. Roosevelt experienced the power of believing. When speaking about the best polio treatment, this world-famous polio victim said, "More important than most therapy is a BELIEF on the patient's part that he will... recover."
- Henry Ford began his business career with the handicaps of poverty, illiteracy, and ignorance. But his undying BELIEF in himself and the potential of the "motor car" led him to overcome these handicaps and become one of the most successful people in the world.
- Neil Armstrong took a giant step for mankind (on the moon) because of the unwavering BELIEF that man could conquer space; because a nation BELIEVED that men in missiles could blast off, land on the moon, take off from the moon, and return safely to Earth; because of the motivational power of BELIEVING that it could be done.

Belief begets behavior and behavior bolsters belief. The combined belief and behavior bring tremendous benefits. Look, for instance, at the organizations that have the best safety and quality performance. Though these industries may have a high potential for accidents or errors that can cause "subquality," they keep that potential under control -- by **believing** in safety and quality and **behaving** accordingly. And the benefits are not only **safety and quality** benefits, but overall **productivity** and **cost** benefits as well, such as:

- Reduced injuries
- Less property damage
- Decreased downtime
- Improved quality
- Better cost control
- Higher productivity
- Increased profitability.

Do *you* believe in SQP? How *strongly* do you believe in SQP? Do you believe that "the right way" to do the job brings these benefits? Do you *support your belief* by your behavior?

Or are you afflicted with all-too-common *negative* beliefs like these?

- **Negative Belief #1:** "Some people are just plain error-prone; they are bound to have accidents and 'subquality' work."

- **Negative Belief #2:** "It's not *their* materials, equipment, and property, so they just don't give a darn."

- **Negative Belief #3:** "We might as well face it -- accidents and 'subquality' incidents occur in clusters."

- **Negative Belief #4:** "Accidents and 'subquality' are basically a matter of luck."

Let's see what the magic of believing has to offer as antidotes for these mind-poisoning negative beliefs.

The masters of "success psychology" have determined how the magic of believing works. People like Dr. Maxwell Maltz, Claude Bristol, Dr. David Schwartz, Paul Meyer, and Dr. Norman Vincent Peale have shown millions through their work and writing how to master the psychology of success/the power of the mind/the magic of believing.

Briefly, here's how it works. Here are five keys to achieving the magic of believing:

1. Set Specific Goals
2. Build Burning Desire and Determination
3. Tap Your Subconscious Power
4. Use the Power of Positive Thinking
5. Be Eternally Persistent

Let's highlight each of these.

1

Set Specific Goals. Mentally crystallize your goals; develop specific plans for reaching them; and set deadlines for goal achievement. Goal-setting is the most important positive action for success. You must decide exactly what you want (and when) -- and put these specifics **in writing**. This is the road map, the plan, the blueprint that guides you to your goals. Writing crystallizes thought, and thought motivates action. When your goals are specific, clear, and vivid, they act as a magnet to attract success.

Hold up your psychological mirror, look yourself straight in the eye, and confront your basic goals. Is one of them "to prevent the killing, disabling, and injuring of human beings"? Is your goal "to prevent errors that cause 'subquality' "? Is it "to protect people, product, process and profits by controlling errors that cause personal injury, occupational illness, property damage, theft, 'subquality,' and other waste"? What **are** your basic goals?

> "Not many of us would trust even a short portion of our lives to an airline pilot who did not have a very precise plan to achieve the goal of getting us from one location to another. Yet we too frequently trust our entire lives or our careers to a safety or quality program that is only buttons, badges, banners, and balloons -- which is both haphazard and unplanned -- when we could, and should, operate with precision."

Goals are the very heart of MBO (Management by Objectives). When you believe in "the **right** way to do the job," you use SQP-MBO. When you **believe**, you set crystal-clear SQP goals; and you capitalize on the motivational power of deadlines and the magic of concentrated effort, clearly directed toward the goals. When you **believe** in SQP, you make sure your goals encompass critical areas of effective safety and quality programs around the world that include (but are not limited to):

- employee selection and placement
- proper job orientation and instruction
- job/task analysis and standard job procedures
- skill training
- rules and practices
- supervisory training
- engineering and purchasing controls
- emergency care
- incident investigation
- planned inspection
- group SPQ meetings.

Because you **believe**, this is how you will **behave**; these are the things you will do; and these are the areas in which you "set specific goals."

2

Build Burning Desire and Determination. A burning desire is the greatest motivator for achievement. The success you achieve depends on the "desire fire" you build. Burning desire is far different from a wish -- it's like a gnawing hunger or a craving thirst. A self-made success who earned his first million before the age of 27, Paul J. Meyer says that to discover the difference between wish and desire you must answer these three questions:

1. What roadblocks must I overcome to achieve my goals?
2. What are the rewards if I attain them?
3. Is it worth it to me?

That third one is the clincher. If your answer is "yes," you'll know you have real desire.

To build desire, train yourself to see not just what **is**, but what **can be**. Use vivid imagination and "picturize" the results. Fan your "desire fire" (or willpower) into flaming motivational force by mental pictures of what **can be**. As a leader, you must do this not only for yourself, but also for those you lead. It is truly the key to creating in them a strong desire to do the things you want them to do. In effect, we take a giant step in gaining the commitment of others as we gain commitment in ourselves. Mentally crystallize and then communicate the benefits of SQP. Paint powerful mental pictures of your goals and their end results. Use the motivational magic in "positive behavior reinforcement." Fan the flames of desire by feeding a human's most basic psychological need: the craving for recognition. Use the power of praise for reinforcing safety and quality behavior. Reap the substantial rewards of recognition and reinforcement. Keep all eyes sharply focused on positive goals and their benefits to build desire and determination.

3

Tap Your Subconscious Power. Everyone occasionally experiences the results of the subconscious mind at work. For instance, you have "a hunch," a sudden inspiration, or a hidden memory suddenly pops into consciousness.

Your subconscious mind records all experiences and thoughts. It stores all sorts of beliefs and aptitudes. You constantly withdraw some things from this "memory bank." Others you may not call upon -- but they are there. Once you learn to tap this submerged source of power, there's no limit to what you can do. **Believing** is the magic that taps this power, since your subconscious mind acts first on your dominant desires -- those intensified by feeling, faith, and belief.

This mental magic is explained by the modern science of **cybernetics**, which deals with **goal-oriented** servomechanisms (similar to those that guide space satellites). In his successful book, *Psychocybernetics*, Dr. Maxwell Maltz explains how the human brain and nervous system serve as a goal-seeking mechanism, operating automatically to reach specific goals. Like other servomechanisms, your subconscious mind is programmed. It is programmed for the goals presented by your dominant mental pictures. Give it negative goals and it works as a "Failure Mechanism." Give it positive goals, and it works as a "Success Mechanism." In other words, WHAT YOU BELIEVE, YOUR MIND FINDS WAYS TO ACHIEVE.

Do you **want** to be committed to an important goal? If you do, you have taken the first giant step that puts you well on the way toward achieving that goal. So keep SQP goals and benefits firmly in mind; and reinforce them in employees' minds with the powers of suggestion, repetition, communication, training, and positive motivation. Indelibly ingrain successful safety, quality, and productivity images into your own nervous system and that of your workers. Use professional management control, for instance, to help human servomechanisms operate as safety-seeking systems. Apply these key aspects of management control:

- **Identification** of work activities to be performed
- **Standards** of performance for the identified work
- **Measurement** of performance
- **Evaluation** of results
- **Counseling** for performance improvement, using both correction of substandard performance and commendation of good safety and quality behavior.

Identifying the work and setting **standards** for it give you the goals toward which people's automatic guidance systems can be programmed. **Measuring** and **evaluating** performance provides the feedback data which, in turn, brings the necessary **counseling** (and corrections) to keep human behavior on course — for safety, quality, **and** productivity.

To achieve this kind of success, boost the belief that safety, quality, and productivity go hand in hand; accentuate positive SQP images; apply the science of cybernetics; and use professional management control. These are ways to tap the enormous subconscious powers which are normally wasted.

4

Use the Power of Positive Thinking. This includes imagination, repetition, and affirmation. Imagination means imagery/thoughts/mental pictures. Basically, what we do depends on our **mental images.** We don't see things as THEY are, but as WE are. As Dr. Maltz says:

> "Your nervous system reacts... to what you **think** or **imagine** to be **true.**"

If you mistake a stick for a snake, your nervous system reacts to the snake. If you see things as hostile, your nervous system reacts to hostility. If you are success-minded, it reacts for success. If you are quality-minded exclusively, it reacts in that one area and may create the cynicism we mentioned earlier, by demonstrating that people are not considered the most important resource we have. If, on the other hand, you are SQP-minded, then safety, quality, and productivity go hand in hand and complement each other. So use your imagination and fill your mind with positive pictures, those mental images of SQP success.

Repetition is another powerful part of positive thinking. Paul Meyer says:

> "An idea heard once and carelessly tossed aside benefits no one! But with REPETITION... REPETITION... REPETITION... ideas become sharp, strong, powerful!"

As successful advertisers know, there is magic in mental images. You get results by creating desired mental pictures. You use carefully spaced repetition to paint mental pictures of the enjoyment, the satisfaction, the benefits to be gained from your product. Likewise, as a member of management, **you** create the mental images that "sell" the right way to do the job by means of:

- personal endorsement, enthusiasm, and example
- communication of policies, procedures, rules, and practices
- programs for proper job orientation and instruction
- talks and tips that reinforce the safety and quality message
- personal contacts, coaching, and counseling.

Once again, it's the unbeatable combination of attitudes and actions, belief and behavior, that indicate safety, quality, and productivity go hand in hand.

"Thoughts, postive or negative, grow stronger when fertilized with constant repetition."

—Dr. David Schwartz
The Magic of Thinking Big

Affirmations are another powerful part of positive thinking to build your own commitment and that of others. In his famous book, *Enthusiasm Makes the Difference,* Dr. Norman Vincent Peale explains affirmations as "self-commands... that you deliberately use to move yourself to desirable action. You merely repeat a verbal self-motivator (often enough) to imprint the words indelibly in your memory... with the **deliberate purpose of getting into action** when the self-motivator flashes from your subconscious to your conscious in time of need."

Positive affirmations are antidotes for mind-poisoning negative beliefs. Remember the four negative beliefs mentioned earlier? Let's take another look at them.

> **Negative Belief #1:** "Some people are just plain error-prone; they are bound to have accidents and 'subquality' work."

This is a negative, defeatist attitude. And as long as we believe it, we will let it come true. But once we decide to accentuate the positive, we will not think of error-prone people -- but of "error repeaters," whose errors are caused. Then we will search for the causes and take positive steps toward their control. It is a well-established concept that **while the majority of accidents and "subquality" incidents** involve substandard acts of people, it is also true that the vast majority (80% or more) can be traced to deficiencies in the **management system (see Exhibits 3 and 4, pages 68 and 69).**

The antidote for "error-prone syndrome" is the actions that accompany positive affirmations like these:

- "I **believe** the errors of 'error repeaters' have specific causes."

- "I **believe** the accidents and 'subquality' work of these repeaters can be controlled by management work such as selection, placement, training, tipping, coaching, and counseling."

Negative Belief #2: "It's not **their** materials, equipment, and property, so they just don't give a darn!"

Believing the worst about people brings out the worst in them. Negative thoughts of this sort are like splinters, festering in the mind. And they do affect your actions. Your beliefs show in your behavior -- and people react to your behavior.

Developing positive safety, quality, and productivity attitudes in the work force requires supervisors who **supervise**/managers who **manage**/leaders who **lead**. As one who believes in "the right way," the SQP way, and one who believes in people, you LEAD people to treat property with respect; you set performance standards and help people meet them; you communicate, educate, and motivate people to the conservation of their own health, the materials and equipment they use, and the products they produce. For the mind-poisoning belief that "people don't give a darn about conserving materials, equipment, property, or product," the antidote is in affirmations like these:

- "I **can** create resource conservation awareness through group communications and personal contacts."
- "I **can** reinforce resource conservation habits by giving recognition to workers who show conservation behavior."
- "I **can** lead the way to conservation of materials, equipment, property, and product."

Negative Belief #3: "We might as well face it -- accidents and 'subquality' occur in clusters."

This is common belief, especially following long safety records. And as long as we continue to believe it, we will continue to have accidents and quality problems in clusters, especially following good records. Mentally, we "let up"; we "give up"; we expect these incidents... and our expectations are fulfilled. This is the **self-fulfilling prophecy** at work. What we think, what we believe, determines what we do. And what we do determines the results we get.

You can make this self-fulfilling prophecy work **for** you in a positive way. The antidote for "giving up" and letting accidents and "subquality" incidents occur in clusters is in the actions that accompany positive affirmations such as these:

- "I **believe** that one accident or 'subquality' incident need not lead to another."

- "I **believe** I can determine basic causes of accidents and 'subquality' incidents and take remedial actions to prevent recurrence."

Negative Belief #4: "Accidents and 'subquality' are basically a matter of luck."

This is the ultimate cop-out, the total defeatist attitude, the rationalization for a "do-nothing" approach. But it simply isn't true! The exact opposite, in fact, is shown to be the case in organizations with a positive belief in prevention and control.

The antidote for the mind-poisoning belief that **accidents and "subquality" are basically a matter of luck is** summarized in affirmations such as these:

- "I **believe** that accidents and 'subquality' incidents are caused and can be controlled."
- "I **can** manage for safety and quality control; I **will** manage for safety and quality control."
- "I **know** that believing in the right way to do the job and **achieving** the right way to do the job go hand in hand."

And here's an affirmation to summarize Key #4: "To achieve total personal commitment and success in my SQP programs, I **will** use the mental forces of imagination, repetition, and affirmation -- the power of positive thinking."

5

Be Eternally Persistent.

"I wish I'd win a fortune."
"I wish people weren't so careless!"
"I wish I had better luck."
"I wish accidents and 'subquality' incidents wouldn't happen."

The largest cemetery in the world is the graveyard of buried wishes. Wishing will **not** make things so -- but **believing** will. Believing and achieving go hand in hand. A sage once said, "the two angels of success are **inspiration** and **perspiration.**" Real faith or belief must be sustained with the sweat of the brow. Turning attitudes into action, beliefs into behavior, takes WORK. Success does not come easily. It takes drive, dedication, and dogged determination, as shown by Thomas Edison. To perfect the incandescent light, he tried not once, not ten times, not even a hundred times -- he tried more than 10,000 times before he met with success.

A **failure** is one who decided to succeed... and **wished**. A **success** is one who decided to succeed... and **worked**. For safety, quality and productivity, this means to train and retrain; to give good group talks week after week after week; in investigating accidents and "subquality" incidents, to keep asking "why" until you find out not only the symptoms, but the **basic** causes; to keep trying to see which SQP management techniques work best for specific situations; to follow up, follow up, follow up. Apply the affirmation, "I will persist until I succeed." In fact, implant those six words firmly into your mind: I WILL PERSIST UNTIL I SUCCEED.

That's the essence of the "magic of believing" and the psychology of personal commitment and success. It reveals the power of beliefs over behavior; of attitudes over accomplishment. We emphasize that **your** beliefs affect your behavior **and** the beliefs and behavior of those around you. Employees tend to believe as their leaders believe. They learn what their leaders believe by observing what their leaders **do.** The major objective of this chapter was to indicate the power of your beliefs upon your own commitment and success, but the overriding importance of your commitment is the fact that it has an enormous influence on those you lead. This is expressed very clearly in a short poem by Forrest H. Kirkpatrick, an industrial operating executive:

"The eye's a better student
    and more willing than the ear;
Fine counsel is confusing
    but example's always clear;
And the best of all the preachers
    are the men who live their creeds;
For to see the good in action
    is what everybody needs.

I can soon learn how to do it
    if you'll let me see it done;
I can watch your hands in action
    but your tongue too fast may run;
And the lectures you deliver
    may be very wise and true;
But I'd rather get my lesson
    by observing what you do.

For I may not understand you
    and the high advice you give;
But there's no misunderstanding
    how you act and how you live..."

As a leader, you have the responsibility, and the golden opportunity, to lead others to **think** SQP, **believe** SQP, and **live** SQP — but you can achieve this goal only through the magic of believing in SQP **yourself**.

CHAPTER 5

Commitment — How Do You Maintain It?

Most organizations have so far to go to **achieve** total program commitment that **maintaining** it is not really the major concern. But for the few that, through outstanding program growth and development, have achieved a level significant enough to say "we made it," the problem of maintenance is very real.

For logical reasons, there is also concern as to "where do we go and what do we do when we get there?" by those who feel they are approaching the point of a significantly high level of program growth and commitment.

The suggestions offered here are based on the combined experience of the professional staff of our Institute, which assists in varying ways with advisement in matters of safety and health to approximately 870 industrial clients. While much more could be said, we believe the heart of the message about how to maintain commitment can be summarized in the following seven points:

1. **Keep reaching, there's always room for growth.**

A few years ago, Dr. Robert H. Hayes, Professor of Business Administration at the Harvard Business School, wrote an article for the *Harvard Business Review* sharing the results of a U.S. productivity study that included a comparison with Japanese industry. In his article, "Why Japanese Factories Work," Professor Hayes stated, "The Japanese have achieved their current level of manufacturing excellence mostly by doing simple things, but doing them very well and slowly improving them all the time." There is the implication in book after book about the Japanese approach to quality that their "inch by inch" improvement process is never-ending.

The book titled *Kaizen* by Masaki Imai (subtitled *The Key to Japan's Competitive Success*) defines the word "KAIZEN" as a gradual, **unending** improvement; doing "little things" better; setting and achieving ever higher standards. "It is KAIZEN," says Mr. Imai, "that is the simple truth behind Japan's economic 'miracle' and the real reason the Japanese have become the masters of 'flexible manufacturing' technology; the ability to adapt manufacturing processes to changing customer and market requirements, and do it fast."

The field of safety and health management is just as dynamic today as the important management function of quality. Hundreds of new chemical substances are available for use in the industrial marketplace today that were not there even last year. The constant influence of technology on machines, equipment, and work processes requires constant vigilance for the need to change in order to keep the workplace safe **and** productive. The challenge to ensure the effectiveness of our human relations efforts to maintain commitment and at the same time to adapt to constant flux is enormous.

Not only must we keep on reaching, bit by bit, to keep attitude and commitment on track with constant program growth and change; but just as important, we must exert this same degree of effort to maintain the ground we have gained. We find that one of the most fruitful activities to make people feel important and to reinforce their belief that management cares about them is to constantly find ways of utilizing their participation in further improvements to the existing safety and quality systems. The "suggestion system approach" is having a great resurgence as one effective tool to accomplish this, combining the satisfaction that comes with participation and seeing one's ideas utilized with special recognition for one's contribution.

Of course, the numerous participative activities used to implement new programs (as described earlier) can now be used to refine and improve them. New team memberships and new, challenging assignments, for example, will frequently do the job. It's more difficult to continue to realize the need for improvement by staff personnel whose role is to continuously assess and advise. Many experienced safety and quality professionals have learned to prime their self-starting, self-motivating pumps by simply taking periodic, serious examinations of what has been accomplished, where things have been and where they are, through a regular walk or drive around the operation -- merely to look for positive change.

This idea evokes the old Southern preacher's prayer:

> "Dear Lord: we ain't what we oughta' be;
> we ain't what we wanta' be;
> and we ain't what we gonna' be;
> but praise da' Lord,
> we sure ain't what we was."

"Live your life each day as you would climb a mountain. An occasional glance toward the summit keeps the goal in mind, but many beautiful scenes are to be observed from each new vantage point. Climb slowly, steadily, enjoying each passing moment, and the view from the summit will serve as a fitting climax for the journey."

—Harold V. Mechert

2. **Periodically update knowledge and skill.**

The efficiency with which a leader executes job analysis, planned observation, or the giving of proper job instruction, is not only directly related to the quality of training he/she received but also to the recency with which that instruction took place. This truth, long known as one of the basic principles of learning, relates in particular to work where structured training is deemed necessary, such as crane or lift operation.

Training specialists and staff advisors in safety and quality recognize that periodic update training is an absolute prerequisite to maintaining the level of competency necessary for performance excellence. They are also fully aware that update training not only involves review but **revision** of knowledge, as well as rejuvenation dealing more specifically with attitude and commitment. Review, revision, and rejuvenation are thus referred to as the "3-R objectives" of update training.

While all three have a strong effect on attitude, the rejuvenation objective requires some special planning and attention. Most progressive companies will have all training sessions opened and closed by a key executive without fail. The opening presents an opportunity for this influential person to relate to and reinforce the organization's mission and policy toward employee safety, health, and quality. The closing of a training session provides an even better opportunity for motivational uplift with commitment reinforcement by an official as earned certificates are presented. If training has truly been well planned and executed, creating the feeling that sincere effort was extended by everyone involved, participants will usually respond with a spirit of appreciation. In this atmosphere, the executive has an opportunity of unusual value for that positive behavior reinforcement so vital to maintaining and strengthening commitment.

3. **Share the fruits of success with others.**

Psychologists and teachers emphasize that there is no experience more rewarding and reinforcing to the individual than sharing ideas and success with others. The greatest value is unquestionably the ego boost that comes through this sharing and meeting of the psychological need to feel important. The value we all would probably rate the highest in evaluating our discussions with others, however, would be the great ideas received from them through this experience.

There are numerous ways that a large number of employees can reap the benefits of this sharing experience. Both internal and external training, for example, provide great opportunity. Training experts today recognize the value and popularity of sessions that offer ample opportunity for people to interact and exchange thoughts and ideas. An increasing number of organizations are forming co-op groups for the very purpose of sharing ideas and experiences that worked well. A wide variety of meetings held by key management personnel with employees are being reported. More and more of these meetings seem to be taking the positive approach of sharing constructive ideas and thoughts, rather than being convened in a manner and atmosphere that stimulates and encourages "bitching, griping, and complaining."

4. **Maintain energy at the source.**

It has been said many times that attitudes, like waterfalls, flow downward. They may certainly be reinforced and sometimes even stimulated by an upward draft or current, but the main source is unquestionably a downward movement. Walter Goldsmith and David Clutterbuck, in their book *The Winning Streak,* state that employees tend to associate the following areas with top management:

- informing employees ahead of time about changes that will affect their jobs
- caring about how employees really feel about their work, and being open and honest in dealing with employees
- giving serious consideration to employee suggestions
- giving supervisors enough authority to get the job done
- making a strong commitment to serving the customer
- having the ability to solve major company problems
- running a socially responsible organization
- providing new services and products required to meet competition
- placing more emphasis on the **quality** than the **quantity** of work.

The way upper management is perceived to deal with these items can determine the attitudes developed toward, and the level of commitment to, the organization.

Numerous management consultants say that one of the biggest myths perpetuated by trainers today is that the front-line supervisor is **the** key person in management. There is no question that the front-line supervisor is a very important person, the "point of control," and in effect a key person in the organization. But let there be no doubt that 99.9% of these key people strive diligently to please their boss. A quotation attributed to Michael Farring states it quite succintly: "Conviction is what the boss thinks."

There is a clear message, then, for the Chief Executive and all those who advise him, including safety and quality professionals. Maintaining proper attitude and commitment of all members of upper management, who have enormous influence on the attitudes of many employees, is probably the most critical aspect of commitment maintenance.

These managers must continue to maintain a leadership role that sets the example and pace for everyone. Experienced staff professionals know there is a tendency for some key executives to "back off" from certain assignments after several exposures. And their judgment may be shared, since many experiences which are dull and repetitive to them may be just as boring and unmotivational to others. This can defeat the very purpose for involvement of key management personnel in those activities that should have enormous motivational values in maintaining commitment. The question most certainly is not whether involvement is vital, but how to put **vitality** into **involvement**.

*"Conviction is
what the boss thinks."*

Priorities of each
Resource in any
Organization
Depend on proper
Utilization and
Commitment of the
Team, and creating
Interest that is
Vital to the
Input
Towards maximum
Yield

5. **Improve your ability to persuade.**

A leader's persuasive ability is so important that this point might well be taken: "Be zealous in the cultivation of your ability to persuade and stimulate people actions and people commitment." We are fully aware that this ability is just as important in the early development and implementation of a program, but it certainly is needed more than ever when things begin to "bog down," when the going gets tough, when you think you're there and you've got to keep interest and commitment to maintain your ground. This capability is absolutely critical in the many interpersonal relations required in every activity, any process step-by-step along the way. There is no doubt that more than most capabilities, it alone can frequently spell the difference between acceptance and rejection, success and failure, in gaining that total commitment.

The January, 1985, issue of *Success* magazine reported on a survey of the Fortune 1000 companies to determine what characterizes a skillful influencer. The forum identified 23 practices exhibited by high performers. These 23 were grouped into three clusters -- planning, conflict, and support. This valuable information appears in Appendix I, and the reader is encouraged to review it.

Professor Raymond J. Wlodkowski, of the Psychology Department of the University of Wisconsin, contributed an article about stimulation in the June, 1985, *Training and Development Journal.* Wlodkowski states: "Stimulation is any change in our perception of experience with our environment that makes us active. We see something more colorful and are attracted to it. We hear something new and listen more carefully to the sound. These are stimulating experiences. Whatever their quality, they get our attention and tend to keep us actively involved. People seek stimulation. There is considerable evidence that sensory stimulation is important to the development and maintenance of normal motivated behavior." Professor Wlodkowski highlights how to maintain attention, build interest, and develop involvement with the conclusion that these stimulation goals can help us better understand the probable outcomes of the motivational strategies we employ.

The authors recently stayed at a Los Angeles hotel that was something less than a 5-Star establishment. As a matter of fact, we wondered whether we should gamble on eating in the dining room, which was even less appealing than the hotel. However, mainly because of convenience, we nevertheless entered the room with considerable trepidation and were promptly seated and greeted by a very pleasant, middle-aged woman, with a smile as broad as her round face.

"My name is Sally," she said, "and how would you like to have a cup of hot coffee made fresh within the past few minutes?" We no sooner nodded our heads than she popped another question as she poured our first cup. "Would you like to have a homemade bun right out of the oven, with pure California butter, to go with your coffee as you take your time to look over the menu?" By now we found our affirmative responses becoming almost automatic, but Sally wasn't finished. "Sir," she said as she placed the warm buns in front of us, "the special today, for $6.95, is a steak prepared with a secret tenderizer by our Polynesian cook, who guarantees you'll like it or double your money back. You can't beat the price, sir, and I personally recommend this unusually tasty, tender steak." Of course, we said, "That's for us."

As we watched Sally dealing with other customers in the dining room, we noticed several of them placing their tips in her hand with some obvious discussion as they left their tables. As a matter of fact, we did the same thing when we finished our meal. We, too, found ourselves commending her on the finest service we had received in a long, long time. Sally was simply practicing these motivational skills involved with stimulation and persuasion.

Sometimes, those who serve in staff positions need to be reminded that their role is to advise, assess, aid, acclaim, activate, affirm, analyze, applaud, appraise, arouse, audit, and awaken; but not to abrade, accost, accuse, agonize, agitate, aggravate, argue, alienate, and anger. The ability to persuade and stimulate people to maintain commitment is of special importance to all leaders.

6. **Don't lose sight of the basics.**

A common danger when a program and its related attitudes and commitments reach a high point is to develop the feeling that basics are no longer an important key to continued success. Some of the frequent queries that arise for leaders and program advisors are: "Where do we go from here?" "What is there to do next?" "What's new that we haven't done?" Then, with what seems to be a complete surprise to everyone, the "bottom falls out" of the program. A sudden increase in rejects, rework, injuries, and illnesses occur. Meetings to discover what is happening seem to be one of the reactions which demonstrate that these events were totally unexpected or unpredictable.

Michael Le Boeuf, in his magnificent little book *The Greatest Management Principle in the World*, has a gem of wisdom that beautifully summarizes what the experts in performance measurement and evaluation say over and over again in different ways: "When you can't understand a problem, go back to the basics and you'll start finding some answers."

Knute Rockne, the famous former football coach at Notre Dame, spoke to one of the largest sales organizations in the world a number of years ago. His message was later said to be one of the most practical and inspiring sales messages ever heard. Here is the heart of it: "At Notre Dame, we have about 300 lads, both varsity and newcomers. They keep practicing fundamentals and keep it up and keep it up and keep it up until these various fundamentals become as natural and subconscious as breathing. Then in the game, they don't have to stop and wonder what to do next when the time comes for quick action. The same thing applies in selling just as well as in football. If you want to be a star in the selling game, you've got to have your fundamentals, the ABC's of your job, so firmly in your mind that they are part of you."

Rockne could very well have been referring to effective management and leadership in general. There is a body of knowledge and skill important to each vital management interest -- Safety, Quality, and Productivity -- that is fundamental to the development of employee capability to do the job right the first time.

We refer to this known body of knowledge and skill as our program and related standards. Safety or quality is frequently defined as "compliance to standards or requirements." The error that causes an accident or "subquality" can likewise be defined as "deviation from a standard or requirement." Point #6, "Don't lose sight of the basics," simply reminds us to maintain a constant, never-ending awareness of the fundamental inputs that control the outputs of our SQP programs. W.E. Deming uses these words: "Create a structure in top management that will push every day on the other points." By doing this effectively, we can be more predictive and far less reactive to the problems that cause accidents and "subquality" losses.

7. Maximize motivation with positive behavior programming.

If there is one thread of continuity through the never-ending proliferation of management texts, it has to be the need to effectively utilize positive behavior reinforcement. Whether it be "situation leadership" or the one-minute whatever, the recurring motivational approach to prompt results and commitment is invariably this relatively simple psychological truth that: behavior is influenced by its effects. A "negative effect" leads to a low probability of behavior repetition, and a "positive effect" leads to a high probability of behavior repetition.

In practical terms, this means that when a specific job behavior results in a negative experience (displeasure, pain, penalty, punishment, frustrated desires), that behavior will tend not to be repeated, and when it results in a positive experience (pleasure, reward, recognition, satisfied desires), the behavior will tend to be repeated.

The proper application of this concept can make the "right way to do the job" so attractive that people will not take the unsafe or unquality way. By balancing the motivational scale with the negative, the wrong way can be made so unattractive that people will be strongly motivated to choose the "right" way.

The values of positive behavior reinforcement are so widely accepted, and richly covered in safety and quality, as well as general management literature, that it suffices to say that we must strive to optimize its effective application in our programs to gain commitment to the "right way to do the job."

Press On

"Nothing in the world can take the place of persistence. Talent will not; nothing is more common than unsuccessful men with talent. Genius will not; unrewarded genius is almost a proverb. Education alone will not; the world is full of educated derelicts. Persistence and determination alone are omnipotent."

8. **Do it again and again and again.**

The modern concepts of behavior reinforcement have been heavily influenced by the animal research of Pavlov and Skinner that utilized repeated positive consequences for the desired behaviors exhibited by the animals.

Aristotle wrote more than 2,000 years ago: "If you would persuade human beings to act as you wish, you must implant motives that lead to consequence-free action, and you must do it again and again."

Philip B. Crosby ends the 14 major points contained in his best-selling book *Quality is Free* with "do it over again."

A recent news release from Taiwan read: "Seven Hundred Love Letters Pay Off.. A young man persisted over two entire years writing 700 letters to a young woman expressing his deep inward feelings of love... This unusual display of 700 persistent individual efforts resulted in the young woman marrying the mailman who made 700 individual personal contacts in making the deliveries.

It is obvious that persistence and determination are vital factors in maintaining momentum and the commitment required to be successful. We also need to be reminded that we should be mentally prepared for periodic setbacks as our program matures. No words seem more applicable to emphasize the need to "hang in there" as the going gets tough than those of Gail Brook Burket, who wrote:

"I do not ask to walk smooth paths,
Nor bear an easy load;
I pray for strength and fortitude
To climb the rock-strewn road.

Give me such courage I can scale
The hardest peaks alone,
And transform every stumbling block
Into a stepping-stone."

CHAPTER 6

Commitment — How Do You Know You Have It?

One of the authors was discussing the subject of commitment with family and friends recently when Nicola, the youngest family member, listening intently to the discussion, volunteered these words which her teacher recited to her class regularly:

> "Good, better, best.
> Never let us rest,
> Until our good is better,
> And our better best."

The author recognized the words immediately and recalled another dedicated teacher using the same words, over fifty-five years ago, to prompt her students to put forth that little extra commitment to gradually move, step-by-step, from good to best. There is really nothing new about the key to Japan's competitive success as defined by the word **Kaizen:**"a gradual, unending improvement; doing 'little things' better; setting and achieving ever higher standards." In retrospect, the author also remembered that his teacher from many years ago would not only have her class recite the little poem, as the child did, but would follow that by asking those willing to put forth that "little extra" to raise their hands.

Philip Crosby, in his book *Quality is Free,* advises the reader to ask for commitments: "Make the audience do something like stand up, or raise their hand, or sign a master commitment chart."

Both of the authors use this psychologically proven approach to send people out of training sessions highly motivated and committed to apply the concepts they have learned. A question is asked, such as: "How many of you agree that too much of the time we do not give clear, effective instructions to assure that our people know 'the right way' to do the job?" This type of question is then followed with another one: "How many of you right now will raise your hand to indicate that you are going to try just a little harder to apply the concepts just taught with your children at home as well as with your fellow workers?" Each question of this type would be followed with appropriate humorous urging to "get those hands up." The resourceful leader can make an experience like this not only enjoyable for everyone, but one that could psychologically lead many people to make a decision that could lead them to a commitment.

In one of the previous chapters we emphasized the value of having a member of upper management open and close training sessions. With a combined experience of over 60 years in human relations, both of the authors know that key executives look to their staff advisors for ways to make their presentations and interfaces with people more effective. The importance of their closing any training session with a call for commitment of the trainees to **use** the knowledge and skill they have gained should be emphasized. Specifically suggest how they might ask for a visible demonstration of the group's feeling and commitment. There are diverse ways this can be done in addition to asking for a show of hands or an audible response to the question. One general manager has had a pledge and commitment to improved management designed and richly imprinted on parchment-style paper. As part of his closing comments for a particular program, he would always announce the very high percentage of his management personnel who have signed this pledge. He would remind them of the leadership role in the industry their organization has held for a significant number of years and highlight the significance that total commitment of everyone has played in maintaining this exemplary position.

He would extend an invitation for them to sign the commitment scroll before leaving and then close on a high motivational note. Again, there is nothing new to the great truth that every salesman is fully aware of: you haven't closed the sale until the commitment to buy has been requested. The advice to close the sale also comes from the Bible:

> "Ask, and it shall be given you; seek, and you shall find; knock, and it shall be opened unto you; for every one that asketh receiveth; and he that seeketh findeth; and to him that knocketh it shall be opened."
>
> Matthew 7:7

Several years ago one of our colleagues was writing a manual about promotional techniques. One of the many interesting items uncovered in his preparatory research was the fact that there is a universal sales formula well known to the majority of sales personnel. It can easily be remembered by use of the acronym AIDA — which stands for:

Attention - get their **attention** by opening your presention with an outstanding positive feature of whatever you are selling.

Interest - build their **interest** by citing at least four additional major values of the item.

Desire - stimulate their **desire** to buy by summarizing the most important personal benefits that will come to them.

Action - close by **asking** for a positive commitment to action.

There is no doubt that one very important means of getting commitment is asking for it. Of course, this is assuming that one has done his/her homework well in preparing for this important moment.

Syndicated columnist and self-made millionaire, Percy Ross, now age 70, gives away thousands of dollars in various forms through his widely acclaimed column, "Thanks a Million," that appears in more than 150 newspapers weekly. Receiving more than 5,000 letters weekly, Ross has a definitive way of deciding who he says yes to and who he turns down. While it may be somewhat oversimplified, his acceptance or rejection largely depends on the way the person asks.

Ross' parents were poor immigrants from Latvia and Russia, and as a child he recognized that his parents' love alone could not provide him with some of the things he dreamed of. At an early age, he decided that wealth could bring him some of those things and he set out at an early goal to become a millionaire. In effect, he asked for the moon and got it. The important thing he learned in the process was that the secret of getting what you want is not just to ask but to know HOW TO ASK. By asking incorrectly, you will very well be refused, but by asking properly, your chances of getting what you want are greatly increased.

He helped his family earn money at an early age through the power of asking people to buy things from him. One of his early challenges came when his father bought him a crate of eggs and requested he go house to house asking neighbors to buy them. This was his first selling job and taught him to sell himself as well as his product. One of the secrets his father taught him was to give the customer an extra egg from time to time. He states that, "You'd think I was giving them an extra dozen." From this experience, he quickly learned one of the big secrets of selling. Give your customers a little more than they expect.

Mr. Ross shares the lessons-in-asking he acquired from his mother and father as a teenager working in a variety of small family businesses. From his mother, he learned to stand tall, be polite, earnest, and persistent when you ask. His dad contributed the positive attitude that characterized a "don't give up" philosophy that prevailed throughout his entire life.

The second dream of multimillionaire Percy Ross was to share his wealth with those less fortunate.

Mr. Ross' 10 rules for asking not only reveals how to "ask for the moon and get it," but advises his readers how to ask for his help and get it.

His rules for asking are:

1. Be sure you know what you want before you ask.
2. Ask the right person.
3. Prepare a good case.
4. Give in order to receive.
5. Overcome the blocks to asking.
6. Ask artfully.
7. Request or invite - never demand or beg.
8. Show respect - the key words are "Please" and "Thank You."
9. Keep asking.
10. Go beyond "me" and 'we."

A detailed explanation of these critical rules for asking can be found in Mr. Ross' book, *"Ask For the Moon and Get It,"* published by G. P. Putnam's Sons, 200 Madison Avenue, New York, NY 10016.

A thorough understanding of them will come with a few hours of most enjoyable reading that could very well change the life of anyone who has hesitated to reach out and ask.

In an earlier section, we indicated that a leader can test a group at the start of a program in order to translate their attitudes (how they think and feel) into behavior that can be evaluated. With oral or written responses, the leader has another indicator of the level of commitment of individuals as well as the group. These and other pieces of valuable information can then be used in developing coaching activities to gain the commitment desired. Likewise, a quiz or test to reflect attitude change as well as knowledge gained can be administered at the end of any program as a new indicator of commitment level.

Of course, the inevitable question must be, "How do I know my own commitment level?" The following questions are designed to assess those actions that a committed person should exhibit. Assess your level of commitment by seeing how many of the twenty you can answer with a firm "yes." Score 5 points for each positive response.

Questions to Assess Commitment

		Yes	No
1.	Do I have a clear mental picture of the mission?	____	____
2.	Are my goals: Specific?	____	____
3.	″ Measurable?	____	____
4.	″ Attainable?	____	____
5.	″ Relevant?	____	____
6.	″ Time-bounded?	____	____
7.	Have I written specific affirmations to support my goals?	____	____
8.	Do I repeat or renew my affirmations daily?	____	____
9.	Do I have a definite plan for progress?	____	____
10.	Do I use mental visualization (imaging) to see what success will look like?	____	____
11.	Do I devote adequate actions to goal attainment?	____	____
12.	Do I really believe that the mission can be met?	____	____
13.	Do I really believe that "I can..."?	____	____
14.	Do I make the necessary sacrifices?	____	____
15.	Do I devote adequate time, effort, and resources to attain the end results that I seek?	____	____
16.	Does time seem to pass quickly when I'm working toward my goals?	____	____
17.	Do I strengthen my desire and determination by working to overcome any barriers and setbacks?	____	____
18.	Does my dedication and enthusiasm for the mission seem to be contagious?	____	____
19.	Do I use the power of positive expectation?	____	____
20.	Do I persist, persist, persist?	____	____
	TOTAL SCORE	____	____

There is some consolation for those who did not score high on this quiz. The fact is that with a little extra thought, consideration, or attention, a decision could have been made in the past that would have altered the answer on the quiz from "no"to "yes." At the end of this section, the authors will provide the means to assist the reader to repond with confidence on future occasions to the high level of commitment that he/she will have.

The experts say you know you are committed when:

- you are steadfastly moved to action toward a vision, a mission, a goal, or a cause.
- you have clear and compelling goals and objectives which sustain all efforts for attainment.
- your mental picture of future success is as clear as present reality.
- you sacrifice immediate satisfactions for the sake of future success.
- your behavior supports your belief in what you are striving for.
- you attract and apply the human, material, and economic resources required to reach your goal.
- the "sweet smell of success" keeps you going in the right direction, on the right track.
- you know you are going to make it, no matter what.
- you maintain motivation in spite of barriers, handicaps, problems, setbacks, disappointments, and failures.
- **you persistently devote all required resources to accomplishment of your mission or goal.**

Frequently there is only a small difference between people who know they have these attributes and those who aren't sure. The enormous power that comes from transforming one's objectives to achieve a goal to written form as a constant reminder and reinforcing mechanism for daily actions has been tapped by one and not by the other. Apply the same powerful psychological values involved with asking others for commitment to asking yourself for it. Complete the personal commitment pledge form that follows (or a copy of one). Place it in a private location that will enable you to review it frequently. We suggest you may want to make one personal commitment pledge related to achieving a business goal and another to achieve one that is quite personal. Childish? Not at all. There is an abundance of research which indicates the power of establishing clearly defined written goals. It's interesting that a recent study of Yale University graduates who have been out of school for 20 years discovered that 87% of the graduates never set any type of goal, 10% could verbally articulate what was important to them, and only 3% operated from a clearly identified, never-changing agenda of written goals. The 87% were dramatically outperformed by the 10%, but not nearly as dramatically as the 10% were outperformed by the 3%.

Whether or not you actually complete and use the specific form provided, success experts around the world would urge you to duplicate the process in some similar fashion. The value to be gained by total commitment to any good cause, whether it be safety and health or product quality — or both — is so enormous that a decision to set and achieve established goals should be an easy one to make.

The significant difference that commitment can make in the support of any cause is well illustrated by this meaningful analogy by Thomas S. Jones, Jr.:

> "Across the fields of yesterday,
> He sometimes comes to me,
> A little lad just back from play,
> The lad I used to be.
> And yet he smiles so wistfully,
> Once he has crept within,
> I wonder if he hoped to see,
> The man I might have been."

Let's not allow a vital desired goal to become an achievement that "might have been" if the proper attitude and commitment had been developed. Don't delay — fill out the personal commitment pledge form that follows. Benefit from the valuable knowledge gained from research about success and goal achievement. If desired, make copies to aid you in achieving more than one important goal.

Personal Commitment Pledge

I, ______________________________ , do believe that...

- persistent pursuit of this goal is vital:

____

____

- attainment of these supportive objectives is critical to accomplishment of my goal:

1. ______________________________________
2. ______________________________________
3. ______________________________________
4. ______________________________________

- I can, I will, devote the resources required to attain my goal.
- I can, I will, make the short-term sacrifices essential to long-term success.
- I can, I will, overcome the barriers, handicaps, problems, setbacks, disappointments, and failures that occur in pursuit of my goal.
- I can, I will, enjoy and share the satisfaction of accomplishment and goal attainment with others who assist in that effort.
- I am committed to achieving my goal.

____________________	____________________
Date of Pledge	Signature

Periodic reviews of pledge and reinforcement:

Dates	Initials	Dates	Initials
________	________	________	________
________	________	________	________
________	________	________	________

APPENDIX 1

23 Proven Persuasion Practices

Planning:

1. Make clear the role each person will play in accomplishing a task.
2. Understand which decisions can be made alone and which cannot.
3. Strive to set team or group goals as well as individual goals.
4. See the value of bringing together people with different opinions.
5. Demonstrate commitment and persistence in achieving goals.
6. Communicate priorities consistently.
7. Insist that group members make every effort to solve problems before taking them to higher management.
8. Keep an open mind about requests to change plans and goals when circumstances seem to warrant change.

Conflict:

1. Encourage reluctant co-workers to express or defend their opinions.
2. Encourage the open airing of problems and differences of opinion.
3. Respond in a non-defensive manner when others disagree.
4. Try to influence others through knowledge and competence rather than official status.
5. Be willing to share one's power in the overall interest of the organization.

6. In conflict situations, look for points of reconciliation rather than differences.
7. Encourage decisions based on logic and the weight of evidence.
8. Encourage the reaching of decisions through a **blending** of ideas.
9. Seek new ways to resolve conflicts.

Support:

1. Behave in a way that inspires trust.
2. Be supportive and helpful to others as they perform their jobs.
3. Encourage innovation and calculated risk-taking in others.
4. Consider others' views according to their logic, not personal preferences.
5. Consider the opinions of others open-mindedly before evaluating them.
6. Evaluate the views of others according to their knowledge and competence rather than their position.

APPENDIX 2

Mission-Oriented Management Activities Related to SQP

1. Broaden the general safety policy to a "Resource Protection Policy" and include concerns for safety of people/equipment/material/product/environment.
2. Promote themes from time to time that apply to productivity and quality as well as safety.
3. Learn to improve your use and application of sound economic principles in the promotion of your program.
4. Broaden incident reporting to include those that could result in "subquality" loss as well as accidents.
5. Broaden investigation standards to include "subquality" incidents as well as those causing injury and/or property damage.
6. Direct your promotional efforts toward those basic causes that result in accidents and "subquality" losses.
7. Teach a loss causation model that covers accidental as well as "subquality" losses.
8. Teach and promote management truths and principles that apply to safety, quality, and productivity.
9. Encourage upper management to promote the use of the words "our SQP program" (representing Safety, Quality, and Productivity).
10. Teach and reinforce problem solving techniques that apply to safety, quality, and productivity.
11. Select basic causes common to accidental and "subquality" losses for assignment to problem solving teams.
12. Minimize paperwork.
13. Make sure your communications with upper management also reflect the efficiencies and

deficiencies that not only relate to safety but also affect quality and productivity.

14. Improve your knowledge of current practices in SQP.
15. Develop a cooperative interdepartmental relationship between SQP staff personnel.
16. Discover and eliminate the barriers to maximize success in any SQP program activity.
17. Make sure all training programs reflect the common values to SQP whenever this correlation is applicable, i.e., Job/Task Analysis and Procedures, Planned Job/Task Observation, clear effective instruction techniques, effective group communication, etc.
18. Emphasize the benefits of **order** that appear frequently in comments about the Japanese methodology of quality control.
19. Teach and promote the other benefits of order, including prevention of accidents and improved customer relations.
20. Teach and promote communication and motivation skills, emphasizing their importance to SQP.
21. Meticulously maintain records of management efficiencies brought about through programs that benefit SQP.
22. Teach and promote the concept that doing things "the right way" includes doing those things that improve SQP.
23. Work diligently to integrate, not separate, safety and quality activities into the right way to do the job.
24. Emphasize the term "loss control" or "control of loss" more frequently in program references in order to emphasize common causes of accidental and "subquality" losses.
25. Do it over and over again!

BIBLIOGRAPHY

Allen, Louis A. *The Principles of Professional Management;* California; Louis A. Allen Associates, Inc.; 1969.

Bennis, Warren and Burt Nanus. *Leaders;* New York; Harper & Row; 1985.

Bird, Frank E., Jr. and Robert G. Loftus. *Loss Control Management;* Georgia; Institute Publishing; 1976.

Bird, Frank E., Jr. and George L. Germain. *Practical Loss Control Leadership;* Georgia; Institute Publishing; 1985.

Blanchard, Kenneth, Ph.D. and Spencer Johnson, M.D. *The One Minute Manager;* New York; Berkley Books; 1981.

Bradford, David L. and Allan R. Cohen. *Managing For Excellence;* New York; John Wiley & Sons; 1984.

Bristol, Claude M. *The Magic of Believing;* New York; Prentice-Hall, Inc.; 1948.

Brown, Arnold and Edith Weiner. *Supermanaging;* New York; New American Library; 1984.

Crocker, Olga L., Syril Charney, and Johnny Sik Leung Chiu. *Quality Circles;* New York; New American Library; 1984.

Crosby, Philip B. *Quality Is Free;* New York; New American Library; 1979.

Delmar, Ken. *Winning Moves;* New York; Warner Books; 1984.

Department of Trade & Industry. *"You Won't Do It!";* London; Moore & Matthes Ltd.; 1985.

Fournies, Ferdinand F. *Coaching For Improved Work Performance;* New York; Van Nostrand Reinhold Company; 1978.

Gallup, George, Jr. and Alec M. Gallup, with William Proctor. *The Great American Success Story;* Illinois; Dow Jones-Irwin; 1986.

Garfield, Charles. *Peak Performers;* New York; Avon Books; 1986.

Gitlow, Howard S. and Shelly J. Gitlow. *The Deming Guide to Quality and Competitive Position.* Foreword by W. Edward Deming. New Jersey; Prentice-Hall, Inc.; 1987.

Goldsmith, Walter and David Clutterbuck. *The Winning Streak;* New York; Penguin Books; 1984.

Green, Mark and John F. Berry. *The Challenge of Hidden Profits;* New York; William Morrow and Company, Inc.; 1985.

Grimaldi, John V. and Rollin H. Simonds. *Safety Management;* Illinois; Richard D. Irwin, Inc.; 1984.

Hawkins, Joyce M. (Compiled by). *The Oxford Paperback Dictionary;* Oxford; Oxford University Press; 1979.

Heinrich, H.W. *Industrial Accident Prevention;* New York; McGraw-Hill Book Company, Inc.; 1950.

Heller, Robert. *The Pocket Manager;* New York; McGraw-Hill Book Company; 1985.

Imai, Masaaki. *Kaizen;* New York; Random House Business Division; 1986.

Insurance Facts; New York; Insurance Information Institute; 1986.

Juran, J. M. *Upper Management and Quality;* 4th Edition; New York; Juran Institute, 1982.

LeBoeuf, Michael. *The Greatest Management Principle in the World;* New York; G.P. Putnam's Sons; 1985.

Maltz, Maxwell. *Psycho-Cybernetics;* New Jersey; Prentice-Hall, Inc.; 1960.

Meyer, Paul. *Professional Salesmanship Program;* Texas; Success Motivation Institute, Inc.

Moss Kanter, Rosabeth. *The Change Masters;* New York; Simon & Schuster, Inc.; 1983.

Murphy, Kevin J. *Effective Listening;* New York; Bantam Books; 1987.

Naisbitt, John and Patricia Aburdene. *Re-inventing the Corporation;* New York; Warner Books, Inc.; 1985.

Peale, Norman Vincent. *Enthusiasm Makes the Difference;* New Jersey; Prentice-Hall, Inc.; 1967.

Peale, Norman Vincent. *The Power of Positive Thinking;* New York; Prentice-Hall, Inc.; 1952.

Peters, Thomas J. and Robert H. Waterman, Jr. *In Search of Excellence;* New York; Warner Books, Inc.; 1982.

Petersen, Dan. *Safety Management - A Human Approach;* New Jersey; Aloray Publisher; 1975.

Rodgers, Buck, with Irv Levey. *Getting the Best Out of Yourself and Others;* New York; Harper & Row; 1987.

Schwartz, David J. *The Magic of Getting What You Want;* New York; William Morrow and Company; 1983.

Schwartz, David J. *The Magic of Thinking Big;* New Jersey; Prentice-Hall, Inc.; 1959.

Uris, Auren. *The Executive Deskbook;* New York; Van Nostrand Reinhold Company; 1976.

Van Fleet, James K. *Power With People;* New York; Parker Publishing Co., Inc.; 1970.

Waitley, Denis. *Seeds of Greatness;* Kingswood; World's Work; 1983.

Waitley, Denis. *The Winner's Edge;* New York; Berkley Books; 1980.

Walton, Mary. *The Deming Management Method.* Foreword by W. Edward Deming. New York; Dodd, Mead & Company; 1986.

About The Authors

FRANK E. BIRD, JR.

Frank E. Bird, Jr., PE, CSP, CRSP, President of the International Loss Control Institute. Adjunct Professor, Georgia State University; former Director of Engineering Services for INA. Eighteen years experience as Supervisor of Safety for the Lukens Steel Company. With 35 years experience in safety and loss control management, he is an internationally known lecturer and writer. His numerous awards and honors include: recognition as Fellow by the American Society of Safety Engineers, Honorary Fellow in the New Zealand Institute of Safety Management, Honorary Life Membership in the Accident Prevention Association of Ontario and the Canadian Society of Safety Engineers, the Public Service Award for his work in occupational safety from the U.S. Secretary of the Interior, and the Distinguished Service Award from the Royal Society for Prevention of Accidents, U.K. He is the author of the books *Management Guide to Loss Control* and *Mine Safety and Loss Control Management;* co-author of the books *Loss Control Management, Damage Control, Practical Loss Control Leadership* and this most recent publication, *Commitment.* He is a contributing author of chapters for the books *Property and Liability Insurance* and *The Industrial Environment - Its Evaluation and Control* by the National Institute of Occupational Safety and Health.

GEORGE L. GERMAIN

George L. Germain, MA, CHCM, ASA, is Executive Consultant for the International Loss Control Insitute. In this capacity, he serves as a researcher and developer of major safety and loss control educational products, leads training conferences, audits and consults on management systems for major corporations. His former positions include: General Manager for the International Safety Academy and Director of Management Development and Training for the Lukens Steel Company. He has been a keynote speaker and conference leader on communication, motivation, safety psychology and loss control management throughout the world. Mr. Germain has published many articles on job analysis, management development, safety communication, and foremanship. Eighteen of his 33 years experience in management development were spent in industrial operations. He is the author of the *Discussion Leader's Guide, Effective Safe Behavior Reinforcement, Effective Job Instruction, Effective Coaching and Tipping* and *Performance Management;* and co-author of the books *Practical Loss Control Leadership, Damage Control,* and *Commitment.*